航 空 無 線 通 信 士

英　語

一般財団法人　情 報 通 信 振 興 会　発行

航空無線通信英語教科書

序文 （preface）

　世界の隅々にまで航空輸送が普及している今日では、各国の言語習慣によるパイロットや管制官などの通信英語の偏りは、運航の安全阻害要因となる可能性を含む。ITU、ICAO はじめ各国の諸規程により、また当事者達の標準化への多大な努力により、航空無線通信英語の世界的標準化が進められてきた。

　我が国における航空無線通信英語の実務は、米国人パイロットにより導入され、その後長年に亘って日本人パイロットおよび関係諸機関によって習熟されてきた。このため米語的、日本語的に偏る傾向があることは否めない。

　本書では、下記の内容を以って、世界的に標準化された航空無線通信英語の学習に努めることとする。

　第1章：定期航空運航および航空無線通信の概要を詳解し、航空無線通信英語への導入とする。

　第2章：ITU、ICAO、電波法、航空法などの規則により国際的に標準化された航空無線通信の基準を示し、それらにかかわる英語の習得に資することとする。

　第3章：ICAO、電波法、航空法およびその他の規定に準拠して、航空無線通信の手順を詳解し、航空無線通信の実務にかかわる英語の習得に資することとする。

　第4章：主として国内定期航空便を例として航空交通管制業務を詳解し、管制業務にかかわる英語の習得に資することとする。

　第5章：航空無線通信の実務を飛行段階に沿って詳解し、通信実務にかかわる英語の習得に資することとする。

第6章：通信・航法の施設・機器について詳解し、それらにかかわる英語
　　　　の習得に資することとする。

(注)　航空無線通信英語が世界規模で標準化されていることの重要性に鑑
　　　み、多くの公的資料を参照して本書を作成した。参照対象資料は
　　　下記の略号により示しておく。

　　　国際電気通信連合　（ITU）　　　　　　国際民間航空機関　（ICAO）

　　　電波法令　（Radio Regulatory Law）　航空法　（CAL）

　　本書は、無線従事者規則第21条第１項第10号の規定により標準教科書
として告示された無線従事者の養成課程用教科書であり、総務省が定め
る無線従事者養成課程の実施要領に基づく項目と内容により編集したも
のです。

目次 (Index)

第3章：航空無線通信の手続き

（Aeronautical Telecommunication Procedures）

4 目次

第4章：航空交通管制（Air Traffic Control）

第5章：航空無線通信の実務（Radio Communication Practice）

第6章：通信・航法装置

（Communication Navigation Systems・Equipment）

---第 1 章---

導　入
Introduction

　航空無線通信英語（ATC English）は、航空機の運航に当たって空地の無線局間において使用される英会話である。言葉は、言わんとする内容が理解されていてはじめて意味を成すものであるから、先ずは「第 1 節：航空機運航」により航空交通管制と運航の概要を説明する。「第 2 節：航空無線通信英語」においては、「会話は作文に始まる」という考えに立って、いくつかの参考事項を述べる。

第 1 節：航空機運航（Flight operation）

　飛行方式には、計器飛行方式（IFR）と有視界飛行方式（VFR）がある。IFR flight の航空機は、飛行中常に ATC の管制下にあるので、パイロットと管制官の間の相互理解は、無線通信会話によって確保される。以下 [1] 飛行と管制、[2] 飛行と ATC 通信により、国内定期運航を例として運航と航空交通管制との関わりを概説する。

[1] 飛行と管制（Flight progress and air traffic control）

表（1）　Flight progress と管制

Pilot	ATC
Flight preparation	
Flight plan の file：pilot と dispatcher で作成した flight plan に両者がサインし、ATC flight plan を ACC に file する。	ACC は、flight plan を review し、flight clearance を発行する。

Clearanceの受領 ： Pilotは、"Delivery"からclearanceを受領し、復唱する。	"Delivery"は、clearance を伝達し、復唱を確認する。
Push back and Taxiing	
Pilotは、"Ground"からpush back clearanceを受領し、地上員にpush backを依頼する。	"Ground"は、push back clearanceをissueする。
Pilotは、"Ground"からtaxi clearanceを受領し、runway手前の停止線までtaxiする。	"Ground"は、taxi clearanceをissueする。"Ground"は、maneuvering areaのtrafficの管制を行う。ASDE又はMLATによるmonitoringを行う。
Takeoff	
Pilotは、"Tower"に離陸準備完了を通報する。	"Tower"は、状況確認後wind data, runway no. その他の必要情報を付してtakeoff clearanceをissueする。
Pilotは、runway no.とtakeoff clearanceを復唱する。所定のprocedureに従い離陸する。	原則として、離陸中のpilotに対する通信は行わない。ASRによるmonitoringを行う。離陸後"Departure"への切り替えを指示する。
Climb	
所定のprocedureに従い上昇する。"Departure"にcontactする。	"Departure"は、出発機にradar serviceを提供する。必要な飛行情報を提供する。
高度14,000ftでQNEをsetする。	Radar serviceを継続する。

"Departure"の指示に従い"ACC"にcontactする。	"Departure"は、出発機に、"ACC"への管制移管の指示を与え、管制を移管する。

En-route	
"ACC"の指示に従い飛行する。	"ACC"は、radar serviceを提供する。 ARSRによるmonitoringを行う。 降下のclearanceを発出する。
降下を開始する。 "ACC"の指示に従い、"Approach"にcontactする。	"Approach"にcontactするよう、指示を与える。管制を移管する。

Approach	
"Approach"にcontactする。 高度14,000フィートでQNHにセットする。Approach clearanceに従い進入を開始する。	Radar serviceを提供する。 Approach clearanceを発出する。
"Approach"の指示に従い、"Tower"にcontactする。	"Tower"にcontactするよう、指示を与える。"Tower"に管制を移管する。

Final Approach	
"Tower"の指示に従い、最終進入を行う。	"Tower"は、pilotに最終進入の誘導を与える。ASRによるmonitoringを行う。

Landing	
"Tower"の着陸許可を復唱する。 所定のprocedureに従い着陸する。	"Tower"は、pilotに着陸許可を与える。Pilotの復唱を確認する。

"Tower"に、滑走路離脱をreportする。	Pilotに、"Ground"へのcontactを指示する。
Taxiing	
"Ground"にcontactする。	"Ground"は、taxi clearanceをissueする。
Taxi clearanceに従って、taxiする。	ASDE又はMLATによりmonitorする。
Parking	
所定の手順で、parking spotにparkingする。	
Completion of Flight	
Pilotは、飛行の終了を所轄のATCに遅滞なく報告する。	Pilotの飛行終了報告を受領する。

（注　空港事務所または出張所が設置されている飛行場では不要。）

［2］飛行とATC通信（Flight progress and ATC communication）

表　(2)　Flight progressとATC通信

Pilot	ATC
Flight Information	
Obtain airport information thru ATIS broadcast.	Tokyo International airport, information Alfa 0600 ILS Z runway 34L approach, landing runway 34L, departure runway 05 and 34R Departure Frequency 126.0 from runway 05, 120.8 from runway 34R Wind 020 degrees 7 knots visibility 9 kilometers, scattered 2,000 feet cumulus, temperature 20, due point 16, QNH 2989 inches, advise you have information Alfa.

Tokyo Delivery, Spaceair 101 Spot 18.	Spaceair 101, cleared to New Chitose airport via Rover 2 A departure flight plan route maintain flight level 170 squawk 5037.
Spaceair 101, cleared to New Chitose airport via Rover 2 A departure flight plan route maintain flight level 170 squawk 5037.	Spaceair 101, your read back is correct, contact Tokyo Ground 121.7 for push back.
PUSH BACK	
Tokyo Ground Spaceair 101 request push back spot 18, information Alfa.	Spaceair 101 push back aproved heading south.
Cleared for push back heading south, Spaceair 101.	
TAXIING	
Tokyo Ground Spaceair 101 request taxi.	Spaceair 101 taxi via Whiskey, Hotel, hold short of Hotel 2.
Taxi via Whiskey, Hotel, hold short of Hotel 2, Spaceair 101 .	
	Spaceair 101 contact Tokyo Ground 118.22.
118.22 Spaceair 101.	
Tokyo Ground, Spaceair 101 on Hotel.	
	Spaceair 101 taxi to runway 34R holding point via Hotel, Charlie.
Taxi via Hotel, Charlie, runway 34R holding point Spaceair 101.	
	Spaceair 101 contact Tokyo Tower 124.35.
124.35 Spaceair 101.	

TAKEOFF	
Tokyo Tower Spaceair 101 on Charlie	
	Spaceair 101 runway 34R line up and wait.
Runway 34R line up and wait, Spaceair 101.	
	Spaceair 101 wind 320 degrees 10 knots runway 34R cleared for takeoff.
Runway 34R cleared for takeoff, Spaceair 101.	
	Spaceair 101 contact departure.
Spaceair 101 good day.	
CLIMB	
Tokyo Departure Spaceair 101 leaving 1600 for FL 170.	Spaceair 101 radar contact.
	Spaceair 101 recleared direct AGRIS climb and maintain FL170.
Recleared direct AGRIS climb and maintain FL170, Spaceair 101.	
	Spaceair 101 contact Tokyo Control 124.1.
124.1 Spaceair 101 good day.	
Tokyo Control Spaceair 101 leaving 12,000 for FL170.	
	Spaceair 101 climb and maintain FL370.
Climb and maintain FL370 Spaceair 101 .	Spaceair 101 contact Tokyo Control 118.9.

118.9 Spaceair 101 good day.	
Tokyo Control Spaceair 101 maintain FL370.	Spaceair 101 Tokyo Control maintain FL370.
Climb and maintain FL370 Spaceair 101.	Spaceair 101 contact Sapporo Control 124.5.
124.5 Spaceair 101 good day.	
Sapporo Control Spaceair 101 maintain FL370.	Spaceair 101, Sapporo Control, proceed direct Naver.
Direct Naver, Spaceair 101 .	
Sapporo Control Spaceair 101 maintain FL370.	Spaceair 101, at pilot's discretion descend and maintain FL210.
At pilot's discretion descend and maintain FL210, Spaceair 101.	
Spaceair 101, leaving FL370 for 210.	
	Spaceair 101, descend and maintain 12,000 QNH 2990, contact Chitose Approach 120.1.
Descend and maintain 12,000 QNH 2990, 120.1. Spaceair 101.	
APPROACH	
Chitose Approach, Spaceair 101, passing FL160 for 12,000, information C.	
	Spaceair 101, descend and maintain 8,000 QNH 2992.
Descend and maintain 8,000 QNH 2992, Spaceair 101.	
	Spaceair 101, fly heading 030 vector to ILS Y runway 01R final approach course, descend and maintain 3,000.

Heading 030 maintain 3,000, Spaceair 101.	
	Spaceair 101, 5miles to YOTEI, cleared for ILS Y runway 01R approach, contact Chitose Tower 118.8.
Cleared for ILS Y 01R approach, 118.8, Spaceair 101.	

LANDING

Chitose Tower, Spaceair 101 approaching YODAI .	
	Spaceair 101, runway 01R cleared to land, wind 350 degrees 5 knots.
Runway 01R cleared to land, Spaceair 101.	
	Spaceair 101, hold short of runway 01L, traffic departing, runway 01L.
Hold short of runway 01L, Spaceair 101.	
	Spaceair 101, cross runway 01L, contact Chitose Ground 121.6.
Clear cross runway 01L, 121.6 Spaceair 101.	
Chitose Ground, Spaceair 101, A4 spot 18.	
	Spaceair 101, taxi via H5, T1, Spot 18.
H5, T1, spot 18, Spaceair 101.	

第 2 節：航空無線通信英語（ATC English）

[1] 作文（Composition）

(1) 文の要素（Element）

何を書くかまたは話すかは、5W1H（who, what, when, where, why, how）に準拠して決める。冗長や重複を避け、簡潔明瞭な表現とすることができる。

① The inspector carefully inspected the turbine section with a bore-scope.

この文では、「carefully」は冗長である。Inspector が注意深く検査するのは当然のことである。

② Pilot が、クリアランスに従って降下を始めるに当たり、「Leaving 370, right now.」といった場合、「right now」の意味は、leaving の中に含まれている。特に強調する場合以外は、不要である。

③ 8,000ft から 3,000ft への descent clearance を受領し、降下していた時の気象状況を述べる場合、「Icing condition continued during descent.」とすると、during descent は、重複である。

(2) 文の構造（Construction）

構文は（S+V+O+C）に準拠して決める。主語と述語を決めて、文の最初におけば、その文の趣旨は表現される。

(3) 構文の要領（Writing a sentence）

5W1H と S+V+O+C を関連付けて文を構築する方法は、実用的である。それには 5W1H を [who does what, when, where, why and how] と考える。これと S+V+O+C を関連付けると、who=S（subject= 主語）、does=V（verb= 動詞）、what=O（object= 目的語）と成って、主たる部分の構文が出来る。補語に相当する部分の [when（時）、where（場所）、why（理由）、how（方法）] は、適宜該当する

10

ところに挿入する。補語に相当する部分が多く構文が複雑になる場合は、文を分けて誤解が生じないようにする。

① 「トランスポンダーのコードは、クリアランスの末尾に与えられている。」を英語表現するには、「transponder code」と「is provided」を最初に述べ、「at the end of the clearance」を加えれば、「Transponder code is provided at the end of the clearance.」となって [S+V+O+C] の形に納まる。

② 「所定の周波数で通信設定が出来なかったので、pilot は、その route に該当する他の周波数で、呼出しを試みた。」前半と後半の二つの文に分ける。前半の文の主語は、「所定の周波数」としてもよいが、後半の文の「pilot」を使ったほうが、全体を平叙文とすることができる。述語は [fail] が使える。したがって前半は、「The pilot failed in communication establishment on the designated frequency.」となる。

後半の文は、「彼は、呼出しを試みた。」で表現する。「He/she attempted to call the station on other frequency appropriate to the route.」前半、後半の文を通して pilot は、同じ局と contact しようとしていることが考えられるので、call の後に the station を挿入する。Call は他動詞である。

(4) 主語 (Subject)

「IFR の航空機は、出発予定時刻の 30 分前までに、飛行計画を file しなければならない。」この文の主語を「IFR の航空機」とすると、動詞の「file する」との間に矛盾が生じる。航空機は飛行計画を file することは出来ないから。この文の主語は、「IFR の航空機の運航者」としなければならない。したがって「An IFR aircraft operator must file the flight plan by 30 minutes before ETD.」となる。

もう一つの例をあげておく。

「航空機は、音速になると主翼の前縁で衝撃波を発する。」「航空機」

を主語とすると、航空機の何が衝撃波を発生するのか、という問題に当面する。これには説明が必要であるが、原文は、その事には言及していない。「衝撃波」を主語として、受動態の文とする方法がある。

(5) 受動態 (Passive)

前出の二つの例は、受動態で表現するに適している。

① The flight plan of an IFR aircraft must be filed by 30 minutes before ETD.

② A shock wave is generated at the wing leading edge of an aircraft when its speed reaches the speed of sound.

(6) 訳語 (Translation word)

訳語の選択を誤ると、誤訳になる。

「One of the objectives of ATC service is to provide information useful for safe and efficient conduct of flight.」において、「conduct」を「管理する」にとるか「実施する」にとるかで、意味が異なってくる。Pilot は、受領した情報を基に飛行を実施するのである。「安全で効率的な飛行の実施に役立つ情報」であって、「安全で効率的な飛行の管理に役立つ情報」では、飛行の実態に合う訳文にはならない。

(7) 言葉の流動性 (Flexibility of meaning)

言葉には、長年の慣習による意味と同時に、流動性による変化もある。実態に合った利用をしないと、意味にずれを生じる可能性がある。「立法の shall」という説明があって、「shall」は、「～しなければならない」と訳されてきた。一方英語習慣としては、「must」、「have to」、「shall」の順序で、禁止の強さが、弱くなる。「shall」には、緩やかな禁止の意味がある。内容の実態に合った訳語を選ぶべきである。

[2] 航空無線通信英会話 （ATC English conversation)

(1) 標準用語 (Standard phraseology)

標準用語や簡略表現は、暗記により習得するが、特定化された条件や内容を持っているので、十分内容を理解することが必要である。

① 標準用語や簡略表現は、指定どおりの条件と指定どおりの意味で使わなければならない。管制から「Contact tower, 118.1」と指示された pilot は、tower との通信設定をしなければならない。一方管制の指示が「Monitor 118.1」であれば、pilot は、受信機に 118.1 をセットして聴守する。pilot は、「contact」と「monitor」のもつ具体的な意味を熟知していなければならない。

② 暗記だけでは、間違っている場合がある。「I have the runway in sight. Request visual approach.」は pilot がよく使う言葉である。ところが pilot の report には、「I have the runway insight.」と書かれていることが非常に多い。耳と口では通用しているが、記述では意味を成さない。

(2) 用語の誤解 (Misunderstanding)

「We lost an engine. We want to come back to Kadena.」と言う米軍 pilot に、「Do you want to go back to Kadena?」と確認を求める日本人 controller。pilot は、「No, I want to come back to Kadena.」、そして controller は、「Do you want to go back to Kadena ?」を繰り返す。年配らしい controller の voice になって、ようやく意思が通じたようであった。

簡明な通信による、確実な世界共通の理解を確保するには、細心の注意と具体的な理解がなければならない。

第 1 章のまとめ英文（The summary of chapter 1）

① Airline flights are operated under IFR. They are operated under the control of ATC and are provided with radar services from taxi-out for departure to taxi-in to the parking spot after arrival.

② Pilots and ATC controllers maintain contact through ATC communications to secure flight safety. The language normally used in ATC communications is English.

③ The use of standard phraseology enhances flight safety.

④ Following principles of "5W1H" and "S+V+O+C" contributes to a concise statement.

⑤ English sentence composition is simple. Determine the subject and verb, and write them at the beginning of the sentence. These two words complete a sentence, for example, [An aircraft has landed.]

第2章

航空無線通信の基準
Standards of Aeronautical Radio Communication

ITU、ICAO の条約及び規則さらには我が国の電波法、航空法には、航空無線通信に関わる世界共通の概念や運用指針が規定されている。この章では、それらの条文などを照会し、航空無線通信の基準に関する英語の習得に資することとする。

The world common concept and policy on the aeronautical radio communications are provided in the provisions of ITU, ICAO, and Radio Laws, and Civil Aviation Laws and Regulations of Japan (CAL and CAR). This chapter intends to promote an understanding of English related to the standards of aeronautical radio communications by referring ITU, other laws and regulations.

第 1 節：国際電気通信連合
(International Telecommunication Union)

[1]　無線周波数と衛星軌道の使用
(The use of radio frequencies and satellite orbits)

The use of radio frequency spectrum, geo-stationary satellite and other satellite orbits.

Member States shall endeavor to limit the number of frequencies and the spectrum used to the minimum essential to provide in a satisfactory manner the necessary services. To that end, they shall endeavor to apply the latest technical advances as soon as possible.

〔用語〕

frequency spectrum 周波数のスペクトル

geo-stationary satellite 静止衛星

Member State 加盟国

the latest technical advance 最新の技術の進歩

〔訳例〕

　無線周波数スペクトルならびに静止衛星および他の衛星軌道の使用

　加盟国は、使用する周波数の数およびそのスペクトル幅を、必要な業務の運用を確保するための最小限に止めるよう努力しなければならない。この目的のために、加盟国は可能な限り速やかに最新の技術の進歩を適用することに努めなければならない。

〔平易な表現〕

Member States shall make efforts to limit the use of frequencies and their spectrum, and the satellite orbits to the minimum to provide the necessary service. Therefore, they should apply the latest technical advances as soon as practicable.	加盟国は使用する周波数、そのスペクトルおよび衛星の軌道は、業務の提供にとって必要な最小限に止めるように努力すべきである。そのため、可能な限り速やかに最新の進歩した技術を適用すべきである。

［2］　周波数と衛星軌道の平等な使用
(Equal rights in use of frequencies and orbits)

　In using frequency bands for radio services, Member States shall bear in mind that radio frequencies and any associated orbits are limited natural resources and that they must be used rationally, efficiently and economically. It should be considered that the countries may have equitable access to those orbits and frequencies.

16

[用語]

　　frequency band 周波数帯

　　radio service 無線通信業務

　　limited natural resource 有限の自然資源

　　equitable access 公平な利用

[訳例]

　　加盟国は、無線通信周波数帯の使用に当たっては、無線周波数およ
び関連衛星軌道が、有限な天然資源であることに留意し、合理的、効
果的かつ経済的に使用しなければならない。さらに各国が、これらの
周波数および軌道を公平に利用できるように考慮しなければならない。

[平易な表現]

Member States shall realize that the radio frequencies and satellite orbits are limited natural resources. They should use frequencies and orbits rationally, efficiently and economically.	加盟国は、無線周波数や衛星軌道が有限な天然資源であることを理解していなければならない。加盟国は、周波数や軌道を合理的、効率的かつ経済的に使用すべきである。

[3]　有害な混信 (Harmful interference)

　All stations, whatever their purpose, must be established and operated in such a manner as not to cause harmful interference to the radio services or communications of other Member States or recognized operating agencies which operate in accordance with the provisions of the Radio Regulations.

〔**用語**〕

harmful interference 有害な混信

radio service 無線通信業務

recognized operating agency 認められた事業体

Radio Regulations 無線通信規則

〔**訳例**〕

　すべての局は、その目的の如何にかかわらず、他の加盟国または認定されかつ無線通信規則の条項に従って実施されている無線業務または通信に対して、有害な混信の原因とならないように設置されかつ運用されなければならない。

〔**平易な表現**〕

All stations must be established and operated not to cause harmful interference to the radio services or communications of other stations that are operating in accordance with ITU-RR.	すべての無線局は、ITU-RRに従って運用されている他の無線業務や通信に対して、有害な混信の原因とならないように開設されまた運用されなければならない。

〔４〕　遭難の呼出しと通報（Distress call and message）

Radio stations shall be obliged to accept, with absolute priority, distress calls and messages regardless of their origin, to reply in the same manner to such messages, and immediately to take such action in regard thereto as may be required.

〔**用語**〕

radio station 無線局　　　　　absolute priority 絶対的優先順位

distress call 遭難呼出し　　　distress message 遭難通報

origin 発信元

18

［訳例］

　無線局は、遭難呼出しおよび遭難通報を、その発信元の如何を問わず、絶対的優先順位をもって受信し、その通報と同様の方法で応答し、かつ必要と思われる措置を直ちにとる義務がある。

［平易な表現］

Radio stations are obliged to accept distress calls and messages with absolute priority regardless of their origin. They are also obliged to reply distress messages and to take action immediately if required.	無線局は、遭難呼出しおよび通報をその発信元の如何を問わず、絶対的優先順位で受信する義務がある。また遭難通報に対して応答することおよびもし必要があれば直ちに措置をとることも義務付けられている。

第 1 節のまとめ英文（The summary of section 1）

① The radio frequencies and the satellite orbits are limited natural resources.

② The use of radio frequencies and satellite orbits should be limited to the minimum.

③ A radio station must not cause harmful interference to communications of other stations.

④ A radio station is obliged to accept distress calls and distress messages with absolute priority, to reply distress messages and to take necessary action.

第２節：国際民間航空機関
(International Civil Aviation Organization)

［１］　通信における規律（The discipline in communication）

In all communications the highest standard of discipline shall be observed at all time. In all situations for which standard radio telephony phraseology is specified, it shall be used.

［用語］

　highest standard of discipline 最高のレベルの規律

　standard radio telephony phraseology 標準無線電話用語

［訳例］

　すべての通信において、常にもっとも厳格な規律が守られていなければならない。無線電話の標準用語が明示されている場合には、標準用語が使用されるべきである。

［平易な表現］

In all communications, an operator should observe the highest standard of discipline at all time.	すべての通信において、操作者は、常に最も高いレベルの規律を守っていなければならない。
An operator should use specified standard radio telephony phraseology whenever it is available.	所定の無線電話標準用語が利用できる場合には、操作者はそれを使うべきである。

［２］　通信設定の責任
　　　　　（Responsibility of communication establishment）

Except as otherwise provided, the responsibility of establishing communication shall rest with the station having traffic to transmit.

20

[用語]

responsibility of establishing communication 通信設定の責任

[訳例]

　別途定められている場合を除き、通信設定の責任は、送信文を持つ局にある。

[平易な表現]

The station having messages to transmit is normally responsible for establishing communication.	通信設定の責任は、通常送信すべき通報を持っている局にある。

[3]　空対空通信　(Air to air communication)

In communications between aircraft stations, the duration of communication shall be controlled by the receiving aircraft station, subject to the intervention of an aeronautical station. If such communications take place on an ATS frequency, prior permission of the aeronautical station must be obtained. Such requests for permission are not required for brief exchanges.

[用語]

aircraft station 航空機局　　　aeronautical station 航空局

ATS frequency 管制の周波数　prior permission 事前の許可

brief exchange 簡単な交信

[訳例]

　航空機局相互間の通信においては、通信の時間は、受信局によって決められる。ただし、この間、航空局による介入はあり得る。もし通信が、ATS 周波数において行われる場合には、航空局の事前の許可を必要とする。この許可は、簡単な交信については必要としない。

［平易な表現］

The receiving aircraft station in air to air communication will control the duration of communication. However, an aeronautical station may intervene.	航空機局相互間の通信においては、通信時間は受信局によって決められる。ただし航空局による介入はあり得る。
If the communication is made on an ATS frequency, prior permission of an aeronautical station is required, except for brief communication.	もし通信が、ATS周波数で行われるのであれば、事前に航空局の許可を必要とする。ただし短時間の通信であれば、許可は必要ない。

［ 4 ］　遭難状態、緊急状態（Distress and urgency conditions）

(1)　Distress is a condition being threatened by serious and/or imminent danger, and requiring immediate assistance.

(2)　Urgency is a condition concerning the safety of an aircraft or other vehicles, or of some person on board or within sight, but one that does not require immediate assistance.

［用語］

　　distress conditions 遭難状態　　　urgency conditions 緊急状態

［訳例］

　遭難状態および緊急状態

　(1)　遭難状態は、重大かつ急迫の危険にさらされ、危急の援助を必要とする状態のことである。

　(2)　緊急状態は、航空機または他の車両、あるいは搭乗者または視界内にある人の安全にかかわる状態、ただし即時の援助を必要とするものではない状態のことである。

［平易な表現］

Distress is a condition that is threatened by serious and/or imminent danger requiring immediate assistance.	遭難状態とは、重大かつ急迫の危険にさらされ、直ちに支援を必要とする状態のことである。
Urgency is a condition that is concerned about the safety of an aircraft or other vehicles, or persons on board or in sight, but does not require immediate assistance.	緊急状態とは、航空機または他の車両あるいは機上の人、または視界内にある人の安全が懸念されるが、直ちに支援を必要とするわけではない状態のことである。

［5］ 標準慣用語句の使用（The use of standard phraseology）

The standard words and phrases prescribed by ICAO procedures shall be used in radiotelephony communications, and their meaning shall be used as described.

［用語］

standard words and phrases 標準用語　ICAO procedures　ICAO 方式
radiotelephony communications 無線電話通信

［訳例］

　無線電話通信においては、ICAO 方式の標準用語が使用され、かつそれらの意味は、そこに示されているように使用されなければならない。

[平易な表現]

The operator should use the standard words and phrases prescribed by ICAO procedures. The meaning of them must be used as described.	オペレーターはICAO方式により規定されている標準用語を使用すべきである。それらの意味は、記述されているとおりに使われなければならない。

第２節のまとめ英文 (The summary of section 2)

① A radio operator should observe the highest standard of discipline.

② The station having messages to transmit is responsible for establishing communication.

③ The receiving station in air to air communication will control the duration of communication. An aeronautical station may intervene in the communication.

④ The air to air communication on an ATS frequency requires prior permission of an aeronautical station except for brief communication.

⑤ Distress is a condition that is threatened by serious or imminent danger and requires immediate assistance.

⑥ Urgency is a condition that is concerned about the safety of an aircraft or persons, but does not require immediate assistance.

⑦ The radio operator should use the ICAO standard phraseology.

第3節：電波法 (Radio Law)

[1] 用語の定義 (Definition of terms)

With respect to interpreting this law and orders issued thereunder, the following definitions shall be used.

(1) Radio wave : electromagnetic waves of frequencies up to 3,000,000MHz.

(2) Radio operator : a person who operates radio equipment or supervises such operation, and holds a license granted by the Minister for Internal Affairs and Communications.

(3) Aeronautical mobile service : a mobile service between aeronautical stations and aircraft stations, or between aircraft stations.

［用語］

electromagnetic wave 電磁波 　　radio operator 無線従事者

radio equipment 無線設備 　　aeronautical mobile service 航空移動業務

aeronautical station 航空局 　　aircraft station 航空機局

［訳例］

　この法令およびこれに基づいて発行された命令の規定の解釈に関しては、次の定義に従うものとする。

(1) 電波：3,000,000MHz 以下の周波数の電磁波をいう。

(2) 無線従事者：無線設備の操作またはその監督を行う者であって、総務大臣の免許を受けたものをいう。

(3) 航空移動業務：航空機局と航空局または航空機局相互間の無線通信業務をいう。

［平易な表現］

Radio waves are electromagnetic waves of frequencies up to 3000GHz.	電波とは、3000GHzまでの周波数の電磁波である。

The radio operator is a person who has an appropriate license and operates radio equipment or controls such operation.	無線従事者とは、該当する免許状を有し、無線装置を操作するかまたは操作の管理を行う人のことである。
Aeronautical mobile service is a mobile service between aeronautical stations and aircraft stations, or between aircraft stations.	航空移動業務とは、航空局と航空機局との間または航空機局相互間の無線通信業務である。

［2］　無線従事者の業務範囲（Scope of radio operator's operation）

The scope of operation of radio equipment or the scope of conducting supervision of operation of radio equipment by an aeronautical radio operator is stipulated in the cabinet order.

Aeronautical radio operator provides:

(1)　Communication operation (except manual Morse telegraphy) by radio equipment installed in aircraft as well as radio equipment of aeronautical stations, aeronautical earth stations and radio navigation for aircraft.

(2)　Technical operation of external adjusting of specified radio equipment.

(3)　Operation allowed to the Amateur Fourth Class Radio Operator.

［用語］

aeronautical radio operator 航空無線通信士

manual Morse telegraphy モールス符号による通信操作

technical operation 技術的操作

external adjusting 外部調整

［訳例］

　航空無線通信士による無線設備の操作および監督の範囲は、政令に関する条例によって規定されている。

　航空無線従事者は、以下のことを実施する。

(1)　航空機に施設する無線設備、ならびに航空局、航空地球局および航空機のための無線航行局の無線設備の通信操作（モールス符号による通信操作を除く）

(2)　無線設備の外部の調整部分の技術操作

(3)　第4級アマチュア無線技士の資格に許可されている操作

［平易な表現］

The scope of radio equipment operation by an aeronautical radio operator is described by the cabinet order.	航空無線通信士による無線装置の操作の範囲は、政令に記述されている。
To communicate by use of radio equipment installed in aircraft, at aeronautical stations and at radio navigation stations.	（航空無線通信士の業務範囲は）航空機、航空局および無線航行局に装備されている無線設備を使って、通信をすることである。
To provide external adjusting of specified equipment.	（航空無線通信士の業務範囲は）無線設備の外部調整をすること。

[3]　講習課程　(Training course)

A training course may be organized satisfying the following requirements (only the selected items are shown below.)

(1)　The approved textbook (standard textbook) shall be used.

(2)　Certain appropriate training subjects are set up satisfying necessary training hours.

The training course is open to anyone who wishes to take, irrespective of age, sex, or nationality.

［用語］

training subject 教育科目　　necessary training hours 必要教育時間

［訳例］

　　以下の要件を満たす場合には講習課程を設けることが出来る（選択項目のみを示す）。

　(1)　当局により認可される教科書（標準教科書）を使用すること。

　(2)　必要授業時間を満たすに適切な授業科目が設定されること。

　　この教育コースは、年齢、性別、または国籍にかかわり無く希望者すべてに開放されている。

［4］　通報の送信方法　（Transmission method of messages）

Messages transmitted shall be spoken clearly and distinctly with pauses between words and phrases. Transmitting speed of messages related to distress, urgency or safety communication shall be such a speed appropriate to be written down by a person receiving the communication.

［用語］

transmitting speed 送信速度　　safety communication 安全通信

［訳例］

　　送信される通報は、はっきり、明瞭にさらに単語や語句の間にポーズ（区切り）を設けて話すこと。遭難、緊急または安全通信にかかわる通報の送信速度は、受信者が書き留めるに適した速さであること。

［平易な表現］

The transmitter shall speak messages clearly and distinctly with pauses between words and phrases.	送信者は、語句の間にポーズをおいて、通報をはっきり、かつ、明瞭に話すべきである。
Transmitting speed of distress, urgency or safety messages shall be appropriate to be written down by the receiver.	遭難・緊急または安全通報の送信速度は、受信者が書き留めるに適した速度であること。

第3節のまとめ英文 (The summary of section 3)

① Radio waves are electromagnetic waves of frequencies up to 3,000,000MHz.

② Aeronautical mobile service is a mobile service between aeronautical stations and aircraft stations, or between aircraft stations.

③ The transmitter should speak messages clearly and distinctly with pauses between words and phrases.

④ Transmitting speed of distress, urgency or safety messages should be appropriate to be written down by the receiver.

第4節：航空法 (Civil Aviation Law)

[1] 航空保安施設 (Air navigation facility)

Air navigation facilities include radio air navigation facilities to support aircraft operations by means of radio waves. Typical facilities are NDB, VOR/DME and TACAN that facilitate aircraft fly the specified route.

［用語］

　radio air navigation facilities 航空保安無線施設

　aircraft operations 航空機運航　　　　　　radio wave 電波

　TACAN（tactical air navigation system）戦術航法装置

　specified route 指定航路

［訳例］

　　航空保安施設には、電波を使用して航空機の運航を支援する航空保
安無線施設が含まれる。代表的な施設としては、航空機が指定のルー
トを容易に飛行できるようにするためのNDB、VOR/DME、および
TACANがある。

[2]　精密進入着陸方式（Precision approach and landing system）

（1）　Instrument landing system（ILS）：ILS provides aircraft on the
final approach with lateral and vertical guidance towards the runway
centerline and to the touchdown　zone, by means of radio waves
transmitted from the ground facilities, the localizer（LLZ）and the
glide slope（GS）equipment.

［用語］

　precision approach and landing system 精密進入着陸方式

　instrument landing system 計器着陸方式　　final approach 最終進入

　lateral guidance 横方向の誘導　　vertical guidance 垂直面の誘導

　touchdown zone 接地帯　　　　　　ground facility 地上施設

　localizer ローカライザー　　　　　glide slope グライドスロープ

［訳例］

　　ILSは、最終進入にいる航空機に対し、滑走路中心線および接地帯に
向かう横方向と垂直面の、地上施設から送信される電波による誘導を
行う。それらはローカライザーおよびグライドスロープである。

(2) Ground controlled approach (GCA)：GCA is a radar approach system consisting of ASR approach and PAR approach. A pilot flies the approach course by following the controller' s instruction during ASR approach phase. A pilot flies the final approach course following GCA controller' s voice instruction based on PAR information, toward the touchdown point of the runway.

〔用語〕

radar approach system レーダー誘導進入方式

approach course 進入経路　　voice instruction 音声による指示

PAR information PARによる情報

〔訳例〕

　GCA は、ASR による進入と PAR による進入とからなるレーダー誘導進入方式である。ASR による進入段階では、パイロットは、管制官の指示に従って進入コースを飛行する。パイロットは、最終進入段階では、PAR の情報による GCA コントローラーの音声による指示に従って、滑走路の接地点に向かって飛行する。

［３］ 無線通信HF, VHF （Radio communication system HF, VHF）

A radio communication system provides direct radio communications between aeronautical stations and aircraft stations by means of HF and VHF communication system within Fukuoka FIR. HF system is applied to long range communication where communication is not covered by VHF system. VHF system is applied to domestic control area and coastal area of the oceanic control.

〔用語〕

radio communication 無線通信　long range communication 長距離通信

domestic control area 国内管制区域　　coastal area 沿岸地域

oceanic control 洋上管制

［訳例］

　　無線通信システムは、HFおよびVHF通信システムにより福岡FIR内において、航空局と航空機局との間に直通の無線通信を提供する。HFシステムは、VHFシステムでカバーできない長距離通信に、VHFシステムは、国内管制区域および洋上管制の沿岸地域に適用される。

[4]　航空交通管制業務 （ATC service）
(1) 業務の目的 (Objects of ATC service)

The purpose of ATC services are to:

① 　prevent collisions between aircraft.

② 　prevent collisions between aircraft and obstructions in the maneuvering area.

③ 　promote and maintain an orderly flow of air traffic.

④ 　provide advice and information useful for safe operation of flight.

⑤ 　inform appropriate organizations regarding aircraft in need of search and rescue aid, and provide support when required.

［用語］

ATC service ATC 業務　　　maneuvering area 走行区域

orderly flow of air traffic 航空交通の秩序ある流れ

appropriate organization 該当する団体

search and rescue aid 捜索救難援助

［訳例］

　　航空交通管制業務の目的は：

　　① 　航空機相互間の衝突を防止すること。

　　② 　走行区域内における航空機と障害物との衝突を防止すること。

　　③ 　航空交通の秩序ある流れを促進しおよび維持すること。

　　④ 　飛行の安全な運行に有用な助言と情報を提供すること。

　　⑤ 　捜索救難援助を必要としている航空機に関する情報を該当する

組織に伝え、必要に応じて支援すること。

(2) 業務の区分 (Classes of ATC service)

ATC service is divided into three parts; area control service, approach control service, and aerodrome control service.

〔用語〕

area control 空域管制　　　approach control 進入管制

aerodrome control 飛行場管制

〔訳例〕

ATC業務は、空域管制業務、進入管制業務及び飛行場管制業務の三つの部分に分けられている。

[5] ATC通信 (ATC communications)

Principal ATC language is English, however, a controller responds in Japanese to a call by a pilot speaking in the Japanese language. The Radio Law requires an appropriate radio operator's license for ATC communication service.

〔用語〕

principal ATC language 主要な ATC 言語　　Radio Law 電波法

radio operator's license 無線従事者免許

〔訳例〕

主要なATC言語は英語であるが、管制官は、パイロットが日本語で呼びかけた場合には日本語で応答する。電波法では、ATC通信業務を行うには該当する無線従事者免許を必要とする。

第4節のまとめ英文 (The summary of section 4)

① Air navigation aids support aircraft operations by means of radio waves. Typical facilities are NDB, VOR/DME and Marker Beacons.

② ILS provides aircraft on final approach with lateral and vertical guidance by means of the localizer (LLZ) and the glide slope (GS).

③ GCA is a radar approach system consisting of ASR approach and PAR approach.

④ A radio communication system consists of HF system and VHF system. HF system is applied to long range communication and VHF system is applied to domestic control area and coastal area of the oceanic control.

⑤ ATC service is divided into three parts: Area control service, Approach control service, and Aerodrome control service.

⑥ Principal ATC language is English, however, a controller responds in Japanese to a call of the Japanese language.

第3章
航空無線通信の手続き
Aeronautical Telecommunication Procedures

　航空無線通信は、ICAO および航空法により設定された標準手順に従って実施される。この章では、ICAO および航空法の規定に準拠して通信手順を概説し、航空無線通信の実施に係わる英語の習得に資することとする。

　The aeronautical radio communication is conducted in accordance with the standard procedures prescribed in ICAO and Civil Aviation regulations. This chapter intends to promote understanding of English related to the operation of aeronautical radio communications, by briefly explaining the ICAO and CAR communication procedures.

第 1 節：通信業務 (Communication Services)

[1] 通信業務の目的 (Object of communication service)

　The object of the aeronautical telecommunication service is to ensure communication and navigation aids necessary for the safety, regularity, and efficiency of flight operations. This service is divided into the following four parts.

(1) aeronautical fixed service

(2) aeronautical mobile service

(3) aeronautical radio navigation service

(4) aeronautical broadcasting service

〔用語〕

　　aeronautical telecommunication 航空無線通信

　　navigation aids 航法援助施設　　flight operations 運航

aeronautical fixed service 航空固定業務

aeronautical mobile service 航空移動業務

aeronautical radio navigation service 航空無線航行業務

aeronautical broadcasting service 航空放送業務

［訳例］

　　航空無線通信業務の目的は、運航の安全、秩序、効率に必要な通信と航法支援を確保することにある。この業務は次の4項目に区分される。

(1)　航空固定業務

(2)　航空移動業務

(3)　航空無線航行業務

(4)　航空放送業務

［２］　通報の区分 (Categories of messages)

The categories of messages handled by the aeronautical mobile service and the order of priority of communications and transmission of messages shall be in accordance with the following table. The priority of NOTAM will depend on its contents and the importance to the aircraft concerned. It may vary from item (3) to (6) of the table.

(1)　Distress calls, distress messages and distress traffic

(2)　Urgency messages, including messages preceded by the medical transports signal

(3)　Communications regarding direction finding

(4)　Flight safety messages

(5)　Meteorological messages

(6)　Flight regularity messages

［用語］

aeronautical mobile service 航空移動業務　order of priority 優先順位

NOTAM ノータム　　distress message 遭難通報

distress traffic 遭難通信　　urgency message 緊急通報

medical transport signal 医療輸送シグナル　direction finding 方向探知

flight safety message 飛行安全通報　meteorological message 気象通報

flight regularity message 飛行正常通報

［訳例］

　航空移動業務において取り扱われる通報の区分およびその通信と通報の送信の優先順位は、以下の表に準拠する。ノータムの優先順位は、その内容および関係する航空機に与える重要度により、下表の（3）から（6）のいずれかになる。

(1)　遭難呼出し、遭難通報、および遭難通信

(2)　医療輸送シグナルを前置きした通報を含む緊急通報

(3)　方向探知に関する通信

(4)　飛行安全通報

(5)　気象通報

(6)　飛行正常通報

第 1 節のまとめ英文（The summary of section 1）

①　The object of aeronautical telecommunication service is to ensure communication and navigation aids required for safety, regularity and efficiency of flight operations.

②　The messages handled by the aeronautical mobile service are categorized and the order of priority is specified. Distress messages have the highest priority.

第 2 節：言語・用語 (Language and Terms)

[1] 言語 (Language)

In general, the air-ground communication shall be conducted in the language normally used by the station on the ground, which is not necessarily the language of the state. For the present time, the English language shall be used and available.

［用語］

air-ground communication 空地通信　　station on the ground 地上局

［訳例］

　一般的に、空地通信は地上局が通常使用している言語で行われるが、必ずしもその地上局の国の言語とは限らない。現在のところ、英語が使用されまた利用可能である。

[2] 標準化 (Standardization)

The pronunciation of the words in the alphabet and numbers may vary according to the language habits of the speakers. In order to eliminate wide variation in pronunciation, transmission of numbers, transmission technique, and standard words and phrases are publicized.

［用語］

language habit 言語習慣　　wide variation 大きな変動

transmission technique 送信要領

standard words and phrases 標準用語

［訳例］

　アルファベットや数字からなる用語の発音の仕方は、話す人の言語習慣によって異なる。発音の大きな相違を避ける目的で、数字の送信、送信要領および標準用語が公表されている。

（1）文字と数字の発音表 (Pronunciation table of letters and numbers)

The Radiotelephony Spelling Alphabet table and the table of the pronunciation of numeral element

無線電話のアルファベット文字

文字	語	国際発音表記法	アルファベットによる表示
A	Alfa	'ælfə	AL FAH
B	Bravo	'bra∶'vou	BRAH VOH
C	Charlie	'tʃa∶li or	CHAR Lee or
		'ʃa∶li	SHAR LEE
D	Delta	'deltə	DELL TAH
E	Echo	'ekou	ECK OH
F	Foxtrot	'fɔkstrɔt	FOKS TROT
G	Golf	gɔlf	GOLF
H	Hotel	hou'tel	HOH TELL
I	India	'indiə	IN DEE AH
J	Juliett	'dʒu∶ljet	JEW LEE ETT
K	Kilo	'ki∶lou	KEY LoH
L	Lima	'li∶ma	LEE MAH
M	Mike	maik	MIKE
N	November	no'vembə	NO VEM BER
O	Oscar	'ɔska	OSS CAH
P	Papa	pə'pa	PAH PAH
Q	Quebec	ke'bek	KEH BECK
R	Romeo	'roumi'o	ROW ME OH
S	Sierra	si'erə	SEE AIR RAH
T	Tango	'tængo	TANG GO
U	Uniform	'ju∶nifɔ∶m or	YOU NEE FORM or
		'u∶nifɔrm	OO NEE FORM
V	Victor	'viktə	VIK TAH
W	Whiskey	'wiski	WISS KEY
X	X-ray	'eks'rei	ECKS RAY
Y	Yankee	'jænki	YANG KEY
Z	Zulu	'zu∶lu∶	ZOO LOO

〔注〕アンダーラインの箇所は、アクセントをつけて発音する。

数字と単位の発音

〔数字〕	〔発音〕	〔数字〕	〔発音〕	〔数値単位〕	〔発音〕
0	∶ZE-RO	5	∶FIFE	Decimal	∶DAY-SEE - MAL
1	∶WUN	6	∶SIX	Hundred	∶HUN-dred

2	: TOO	7	: SEV-en	Thousand	: TOU - SAND
3	: TREE	8	: AIT		
4	: FOW-er	9	: NIN-er		

〔注〕大文字で表示した音節は、アクセントをつけて発音する。

(2) 数字の送信 (Transmission of numbers)

　　All numbers, except as prescribed apart, shall be transmitted by pronouncing each digit separately.

〔用語〕

　digit 数字

〔訳例〕

別途定める数字以外は、数はすべて個々の数字ごとに送信する。

科目例	項目例	送信例
Aircraft call sign	Space Air 101	Spaceair WUN ZERO WUN
Flight levels	FL 180	flight level WUN AIT ZERO
Headings	180 degrees	heading WUN AIT ZERO
Wind direction and speed	200 degrees 70 knots	wind TOO ZERO ZERO degrees SEVen ZERO knots
transponder codes	2400	squawk TOO FOWer ZERO ZERO
Runway	27	Runway TOO SEVen
Altimeter setting	1010	WUN ZERO WUN ZERO
	29.80	QNH TOO NINer AIT ZERO

(3) 数字の発音 (Pronunciation of numbers)

　　All numbers used in the transmission of altitude, cloud height, visibility and runway visual range （RVR）, which contain whole hundreds and whole thousands, shall be transmitted followed by the word HUNDRED or THOUSAND. Combination of thousand

and whole hundreds shall be transmitted by pronouncing each digit in the number of thousands followed by the word THOUSAND followed by the number of hundreds followed by the word HUNDRED.

［用語］

cloud height 雲高　　visibility 視程

runway visual range（RVR）滑走路視距離　　whole hundreds 百単位の

Altitude	800	AIT HUNDred
	3,400	TREE TOUSAND FOWer HUNdred
	12,000	WUN TOO TOUSAND
Cloud height	2,200	TOO TOUSAND TOO HUNdred
Visibility	1,000	visibility WUN TOUSAND
	700	visibility SEVen HUNdred
Runway visual range	600	RVR SIX HUNDred
	1,700	RVR WUN TOUSAND SEVen HUNDred

［訳例］

　高度、雲高、視程およびRVRの送信においては、百および千の位の時には"HUNDRED"または"THOUSAND"をつけて送信する。千の位と百の位を含む場合には、千の位の数字を発音しそれに"THOUSAND"を付け、さらに百の位の数字と"HUNDRED"を付加して送信する。

（4）小数点を含む数字（Numbers containing a decimal point）

　Numbers containing a decimal point shall be transmitted indicating the decimal point by the word DECIMAL.

Radio frequencies

118.0　　WUN WUN AIT DAYSEEMAL ZERO

118.12　　WUN WUN AIT DAYSEEMAL WUN TOO

〔用語〕

decimal point 小数点

〔訳例〕

　小数点を含む数字は、DECIMALの語を用いて小数点を表示して送信する。

(5) 時刻 (Time)

　When transmitting time, only the minutes are normally required. Each digit of the minutes shall be pronounced separately. However, the hour shall be included when possibility of confusion is likely to occur. INDIA or ZULU is suffixed to the digits of minutes to indicate the time in the local time or in UTC respectively.

0930	TREE ZERO or ZERO NINer TREE ZERO
1645	WUN SIX FOWer FIFE
0930I	ZERO NINer TREE ZERO INDIA
0930Z	ZERO NINer TREE ZERO ZULU

〔用語〕

possibility of confusion 混乱の可能性

local time 地方時　　UTC 世界協定時

〔訳例〕

　時刻の送信にあたっては、通常"分"のみとする。"分"の各数字は1字ずつ発音する。ただし、混乱が起こる可能性がある場合には、"時間"を含む。地方時または世界協定時の表示をする場合には、"INDIA"または"ZULU" をそれぞれの"分"の数字の後におく

(6) 数字の確認 (Verification of numbers)

　When it is desired to verify the accurate reception of numbers, the person transmitting the message shall request the preson receiving to read back the numbers.

［用語］

accurate reception 正確な受信　　person transmitting 送信者

person receiving 受信者　　　　　read back 復唱

［訳例］

　数字が正確に受信されていることを確かめたい時には、通報の送信者は受信者に数字の復唱を要求すべきである。

[3] 送信要領 (Transmitting technique)

(1) 簡潔な送信 (Concise transmission)

Transmission shall be made concisely. For this purpose, air crew and ground personnel should slowly enunciate each word clearly and distinctly to allow for the writing process. A slight pause preceding and following numerals makes the receiver easier to understand.

［用語］

concise transmission 簡潔な送信　　ground personnel 地上員

enunciate 明確に発音する　　　　　writing process 書取り

slight pause 少しのポーズ

［訳例］

　送信は簡潔に行うこと。それ故、乗員および地上員は、各語を明瞭明白にかつ書き取りが出来るようにゆっくりと明確に発音すべきである。数字の前後に若干の間を置くことは、受信者の容易な理解につながる。

(2) 長文通報の送信 (Long message transmission)

The transmission of long messages shall be interrupted momentarily from time to time to permit the transmitting operator to confirm that the frequency in use is clear and, if necessary, to permit the receiving operator to request repetition of parts not received.

［用語］

long message 長文の通報　transmitting operator 送信者

frequency in use 使用中の周波数

receiving operator 受信者　parts not received 受信されなかった部分

［訳例］

長文通報の送信は、送信者が使用中の周波数が正常であることを確かめ、またもし必要であれば、受信者が受信できなかった部分の反復を要求することが出来るように、時々少しの間、間をおいて行うべきである。

(3) 標準語句の使用 (Use of standard phraseologies)

Message transmission shall be conducted by use of standard phraseologies wherever they are available in ICAO procedures. The following is ICAO standard phraseologies. They shall be used in radio communications, and they shall have the meanings as described.

［用語］

standard phraseology 標準語句　message transmission 通報の送信

ICAO procedure ICAO 方式　radio communication 無線通信

［訳例］

通報の送信に当たっては、ICAO方式の標準語句が利用可能な場合にはそれらを使用すべきである。以下はICAOの標準用語句である。無線通信に当たってはこれらを使用し、その意味も記述されているとおりに活用されるべきである。

[Words and Phrases]	[Meaning]
ACKNOWLEDGE	"Let me know that you have received and understood this message."
	この通報を受信しかつ理解したことを返信してください。

AFFIRM	"Yes." その通りです。
APPROVED	"Permission for proposed action granted." 申し入れ事項を許可します。
BREAK	"I hereby indicate the separation between portions of the message." これより、通報の各部分の区切りを示します。
BREAK BREAK	"I hereby indicate the separation between messages transmitted to different aircraft in a very busy environment." 多忙な状況にあるので、他の航空機との通報の間に区切りを示します。（続けて他局と交信しますので、受信証の送信は不要。）
CANCEL	"Annul the previously transmitted clearance." 前に送信したクリアランスを無効にします。
CHECK	"Examine a system or procedure." システムまたは手順をチェックしてください。
CLEARED	"Authorized to proceed under the conditions specified." 所定の条件の下での実行を承認します。
CONFIRM	"Have I correctly received the following ~?" or "Did you correctly receive this message?" 次の～についてこちらの受信は間違っていませんか？またはそちらではこの通報を正しく受信しましたか？
CONTACT	"Establish radio contact with ~." ～と通信設定してください。
CORRECT	"That is correct." その通りです。

CORRECTION "An error has been made in this transmission (or message indicated). The correct version is ~."

この送信（または指摘の通報）に誤りがあります。正しいものは～です。

DISREGARD "Consider that transmission as not sent."

送信は送られなかったものとみなしてください。

GO AHEAD "Proceed with your message."

通報を送信してください。

HOW DO YOU READ? "What is the readability of my transmission?"

当方の送信の感明度は如何ですか？

I SAY AGAIN "I repeat for clarity or emphasis."

はっきりさせるためまたは強調のために反復します。

MONITOR "Listen out on (frequency) ."

（周波数）を聴守してください。

NEGATIVE "No" or "Permission not granted." or "That is not correct."

"いいえ"、"許可できません" または "それは間違いです。"

OVER "My transmission is ended, and I expect a response from you."

Note: Not normally used in VHF communications.

この送信は終了しました。そちらからの応答を待ちます。

（注）VHF 通信では通常使用されない。

OUT "This exchange of transmissions is ended and

no response is expected." Note: Not normally used in VHF communications.

この交信は終わりました。応答は不要です。

（注）VHF 通信では通常使用されない。

READ BACK "Repeat all, or the specified part, of this message back to me exactly as received."

通報の全文または指定の部分を受信した通りに反復してください。

RECLEARED "A change has been made to your last clearance and this new clearance supersedes your previous clearance or part thereof."

先ほどのクリアランスに変更があります。この新しいクリアランスが先ほどのクリアランスまたはその一部に取って代わります。

REPORT "Pass me the following information."

次の情報を通報してください。

REQUEST "I should like to know ~." or "I wish to obtain ~."

"〜を要求します。" または "〜を入手したい。"

ROGER "I have received all of your last transmission." Note: Under no circumstances to be used in reply to a question requiring "READ BACK" or a direct answer in the affirmative (AFFIRM) or negative (NEGATIVE).

そちらの送信のすべてを受信しています。

（注）どのような条件下でも READ BACK を要求する質問に対する回答または承諾（AFFIRM）または否定（NEGATIVE）の直接の回答として使ってはならない。

SAY AGAIN	"Repeat all or the following part, of your last transmission." 先ほどの送信の全部または以下の部分を再送信してください。
SPEAK SLOWER	"Reduce your rate of speech." 送信速度を落としてください。
STANDBY	"Wait and I will call you." 待ってください。こちらから呼び出します。
VERIFY	"Check and confirm with originator." 発信者に確認して照合してください。（確認してください。）
WILCO	"I understand your message and will comply with it." 通報を了解しました。それに従います。
WORDS TWICE	① As a request："Communication is difficult. Please send every word, or group of words, twice." 要求の場合：通信が困難です。それぞれの単語または語句を 2 度送信してください。 ② As information："Since communication is difficult, every word, or group of words, in this message will be sent twice." 情報提供の場合：　通信困難のため、この通報中の単語または語句を 2 度ずつ送信します。

第 2 節のまとめ英文 (The summary of section 2)

① The air-ground communication uses the English language.

② The language habits cause variation in pronunciation. Use of the

ICAO standard phraseology in radio communications is required to eliminate wide variation.

③　Transmission should be made concisely. Each word should be pronounced clearly and distinctly with a slow speaking rate.

第3節：通信要領（Communication Procedures）

[1]　呼出符号（Call signs）
（1）航空局（Aeronautical stations）

Aeronautical stations are identified by： (1)　the name of the location；and　(2)　the unit or service applicable.

〔用語〕

unit 機関

〔訳例〕

航空局は、その所在地名と、該当する機関名または業務名からなるコールサインによって識別される。

Unit/service	Call sign suffix
Area control center	CONTROL
Approach control	APPROACH
Approach control radar arrivals	ARRIVAL
Approach control radar departure	DEPARTURE
Aerodrome control	TOWER
Radar (in general)	RADAR
Surface movement control	GROUND
Clearance delivery	DELIVERY
Company dispatch	DISPATCH

(2) 航空機局 (Aircraft stations)

There are three types of aircraft call sign as indicated below.

Type A ： the corresponding to the registration marking of the aircraft.

Aircraft registration marking of this country consists of "JA" (nationality) , and four-digit numerals or three-digit numerals with one alphabetic letter. Example：JA6001, JA731J

Type B ： the telephony designator of the aircraft operating agency, followed by the last four characters of the registration marking of the aircraft. Example：Spaceair 6001

Type B call sign is applicable to the non-revenue flight such as a test flight, or a training flight.

Type C ： the telephony designator of the aircraft operating agency, followed by the flight identification. Example: Spaceair 101

Type C call sign is applicable to the revenue flights, and no abbreviated form is allowed.

〔用語〕

registration marking　登録記号　　four-digit numerals 4 桁の数字

telephony designator　無線電話識別

aircraft operating agency　航空機運航機関

non-revenue flight　非有償飛行　　fligfht identification 飛行便名

revenue flight　有償飛行　　　　abbreviated form 簡略形式

〔訳例〕

航空機のコールサインには、下記に示す3通りのタイプがある。

タイプA：航空機の登録記号に対応する記号。我が国での登録記号は、国籍記号（JA）と4桁の数字または3桁の数字と1つのアルファベット文字から成る。「例：JA 6001、JA731J」

タイプB：航空機運航機関の無線電話識別とそれに続く登録記号の最後

の4桁の記号とから成る。「例：Spaceair 6001」

タイプBコールサインは、試験飛行、訓練飛行などの非有償飛行に適用される。

タイプC：航空機運航機関の無線電話識別とそれに続くフライトナンバーとから成る。「例：Spaceair 101」

タイプCコールサインは、有償飛行に適用される、簡略化は許されない。

(3) 航空機局コールサインの変更（A change of an aircraft call sign）

An aircraft shall not change the type of its radiotelephony call sign during flight, except temporarily on the instruction of an air traffic control unit in the interests of safety.

［訳例］

航空機局は、飛行中にその無線電話の呼出符号の形式を変更してはならない。ただし、一時的に安全のため航空交通管制所の指示に基づくものを除く。

［2］ 通信要領（Communication procedures）

Except for reasons of safety, no transmission shall be directed to an aircraft during takeoff, during the last part of the final approach or during the landing roll.

［用語］

takeoff 離陸　　final approach 最終進入　　landing roll 着陸滑走

［訳例］

安全上必要な場合を除き、離陸中、最終進入の最終段階にあるまたは着陸滑走中の航空機に対して、送信してはならない。

(1) 通信設定（Communication establishment）

Establishment of communication shall be initiated by the station having messages to transmit, using the full call signs.

〔用語〕

establishment of communication 通信設定

message to transmit 送信すべき通報

〔訳例〕

　通信設定は、送信すべき通報を持っている局が所定のコールサイン
を使って開始する。

（2）　不確実なコールサイン（Uncertainty of the station call sign）

　When a station is uncertain of the call sign of the calling station,
it should reply by transmitting the following：

　Station calling,（receiving station call sign）, say again your
call sign.

〔用語〕

calling station 呼出しをしている局

〔訳例〕

　呼出しをしている局のコールサインが不確実な場合には、下記のよ
うに送信することによって答える。

　Station calling,（receiving station call sign）, say again your call
sign.

（3）　一括呼出し（Transmitting information to "all stations"）

　Stations having a requirement to transmit information to all
stations should preface the transmission by the general call ALL
STATIONS, followed by the call sign of the calling station. No
reply is expected to ALL STATIONS calls unless stations are
requested to reply.

〔用語〕

general call 総括呼出し

〔訳例〕

　一括送信を必要とする局は、送信に総括呼出しのALL STATIONSを
前置し、その後に自局のコールサインを続ける。一括呼出しに対して

は、要求されない限り回答は必要としない。

(4) 空対空通信 (Interpilot communication)

Interpilot air-to-air communication shall be established on the appropriate frequency by either a direct call to a specific aircraft station or a general call. The initial call should include an indication of air-to-air identification "INTERPILOT".

The following is the example of initial call of interpilot communication：

① Spaceair 101, Marsair 201, INTERPILOT, DO YOU READ

② Any aircraft vicinity of 40 north 140 east, Spaceair 101, INTERPILOT, 128.95, over

〔用語〕

interpilot air-to-air communication　パイロット相互間の空対空通信

direct call 直接呼出し　　　　general call 総括呼出し

initial call 最初の呼出し　　　air-to-air identification 空対空の識別

〔訳例〕

飛行中の航空機相互間の通信は、該当する周波数で、特定の航空機に対する直接呼出しまたは総括呼出しで行う。最初の呼出しには"空対空"を意味する"INTERPILOT"を含むこと。以下は航空機局間通信の最初の呼出しの例である。

① Spaceair 101, Marsair 201, INTERPILOT, DO YOU READ

② Any aircraft vicinity of 40 north 140 east, Spaceair 101, INTERPILOT, 128.95, over

第3節のまとめ英文 (The summary of section 3)

① Call signs

An aeronautical station and an aircraft station are identified by their specific call sign, which is used for establishment and exchange of communication.

② Interpilot air-to-air communication

The initial call of air-to-air interpilot communication should include the word INTERPILOT.

第4節：通信の実行 (Communication Practices)

[１] 交信手順 (Exchange of communications)
(1) 通信設定 (Establishment of contact)

Establishment of contact is made by conduct of the following calling procedure and the reply procedure.

① Calling procedure：

Designation of the station called ⇒ designation of the station calling

② Reply procedure：

Designation of the station called ⇒ designation of the answering station ⇒ invitation to proceed with transmission

［用語］

establishment of contact 通信設定　　calling procedure 呼出し手順
station called 呼び出される局　　station calling 呼び出している局
reply procedure 応答手順　　answering station 応答する局

［訳例］

通信設定は下記の呼出し及び応答手順の実施により行われる。

① 呼出しの手順

呼出しを受ける局のコールサイン⇒呼出しをする局のコールサイン

② 応答手順

　　　呼出しをした局のコールサイン⇒応答する局のコールサイン⇒
　　送信への誘い (GO AHEAD)

(2) テストの手順 (Test procedures)

An aircraft station may request an aeronautical station a conduct
of equipment test by sending the test transmission forms as
described below：

① The identification of the station being called

② The aircraft identification

③ The words RADIO CHECK

④ The frequency being used

The reply to the test transmission should be as follows ：

① The identification of the aircraft

② The identification of the aeronautical station replying

③ Information regarding the reliability of the aircraft
transmission

Reply should be made by use of the following readability scale：

1. Unreadable

2. Readable now and then

3. Readable but with difficulty

4. Readable

5. Perfectly readable

［用語］

equipment test 装備のチェック

test transmission form テスト送信フォーム

readability scale 感度スケール

［訳例］

　　航空機局は、航空局に対して下記のテスト送信フォームを送信する
ことによって、装備のテストの実施を要求することができる。

① 呼出しを受ける局のコールサイン

② 航空機のコールサイン

③ RADIO CHECKの語

④ 使用している周波数

テスト送信に対する応答は、以下のとおりとする。

① 航空機のコールサイン

② 応答する航空局のコールサイン

③ 航空機の送信の感度に関する情報

感度は以下のスケールに従うこと。

1. 聞き取り不能

2. 時々聞き取り可能

3. 困難を伴うが聞き取り可能

4. 聞き取り可能

5. 十分聞き取り可能

(3) 受信証 (Acknowledgement of receipt)

An aircraft station shall acknowledge receipt of important ATC messages by reading them back and terminating the read-back by its radio call sign. ATC clearances, instructions and information requiring read-back are separately specified. If position report and other information such as weather report are received in the same message, the information should be acknowledged with the words such as "WEATHER RECEIVED" after the position report has been read back.

〔用語〕

acknowledgement of receipt 受信証

important ATC message 重要な ATC 通報

read back 復唱　　　　　　　　separately specified 別途指定の

position report 位置通報　　　weather report 気象報告

56

[訳例]

　航空機局は、重要なATC通報については自機のコールサインを付して復唱を完了することによって、その受信の確認をしなければならない。復唱を必要とするATCのクリアランス、指示および情報は、別途指定される。もし位置通報と他の情報例えば気象報告が同じ通報の中で受信された時には、その情報の受信は、位置通報の復唱が終了した後に、例えば WEATHER RECEIVED のような語を用いて確認する。

（4）　会話の終了（End of conversation）

A radio communication conversation shall be terminated by the receiving station using its own call sign.

[用語]

radio communication conversation 無線通信会話
receiving station 受信している局

[訳例]

　無線通信会話は、受信局がそのコールサインを送信することによって終了する。

[2]　訂正と反復送信（Corrections and repetitions）
（1）　訂正（Correction）

When an error has been made in transmission, the word CORRECTION shall be spoken, the last correct part repeated, and then the correct version transmitted.

[用語]

last correct part 最後の正しい部分　　correct version 正しい文

[訳例]

　送信中に間違いを生じた場合には、CORRECTIONという語を送信し、最後の正しい部分を反復送信し、それから正しい文を送信すること。

(2)　全文訂正 (Correction of an entire message)

If a correction can best be made by repeating the entire message, the phrase CORRECTION, I SAY AGAIN shall be spoken before transmitting the message at second time.

［用語］

entire message 通報の全部

［訳例］

もし通報の全部を繰り返し送信することで訂正することが最も良い訂正の方法である場合には、CORRECTION, I SAY AGAINという語を二度目の送信の前に送信すること。

(3)　強調 (Emphasis)

① When an operator transmitting a message considers that reception is likely to be difficult, he/she shall transmit the important element of the message twice.

② If the receiving operator is in doubt of the message received, he/she shall request repetition of the message.

［用語］

an operator transmitting a message 通報を送信中のオペレーター

important element 重要な要素

receiving operator 受信側のオペレーター

［訳例］

① 送信中のオペレーターは、受信に困難があるようだと思った時には、通報の中の重要な部分を二度送信すること。

② 受信側のオペレーターは、受信した通報に疑いを持った場合には、通報の再送信を要求すること。

(4)　復唱の訂正 (Correction of read-back)

If an error has been found in a read-back, the person receiving the read-back shall transmit the words NEGATIVE I SAY AGAIN at the conclusion of read-back followed by the correct version.

〔用語〕

conclusion 終わり　　followed by ～に続けて

〔訳例〕

　もし復唱の中に間違いを見つけた場合には、復唱の受信者は、復唱の終了時にNEGATIVE I SAY AGAINといって、その後に正しい通報を引き続き送信する。

[3]　通信の聴守（Communication watch）

（1）　航空機局による通信の聴守
　　　　　（Communication watch by an aircraft station）

　Aircraft station shall maintain communication watch during flight except for reasons of safety.

〔用語〕

communication watch　通信の聴守　　reasons of safety 安全上の理由

〔訳例〕

　航空機局は飛行中、安全にかかわる要件の場合を除き常時通信を聴守する。

（2）　航空機局による非常用周波数 121.5MHz の聴守
　　　　　（121.5MHz watch by an aircraft station）

　Aircraft on long over-water flights, or on flights over designated areas over which the carriage of an emergency locator transmitter(ELT) is required, shall continuously guard the VHF emergency frequency 121.5MHz, except for those periods when aircraft are carrying out communications on other VHF channels or when airborne equipment limitations or cockpit duties do not permit simultaneous guarding of two channels.

［用語］

long over-water flight 長距離洋上飛行　　designated area 指定地域

carriage of an emergency locator transmitter 非常用位置指示送信機の
備付け

VHF emergency frequency　VHF非常用周波数

［訳例］

　長距離洋上飛行または非常用位置指示送信機(ELT)の備付けが要求される指定された地域の飛行に従事する航空機は、VHF非常用周波数121.5MHzを常時聴守しなければならない。ただし、航空機が他のVHFチャンネルで通信に従事している時間又は搭載設備の制限若しくは操縦席の任務が二つのチャンネルの同時聴守を許さない場合を除く。

［4］　通信不能 (Communication failure)

(1)　航空機局の措置 (Procedures by an aircraft station)

　When an aircraft station fails to establish contact with the aeronautical station, it shall attempt to establish contact on another appropriate frequency. If this attempt fails, the aircraft station shall attempt to establish communication with other aircraft station or aeronautical station.

［用語］

establish contact 連絡設定する

another appropriate frequency 他の適切な周波数

［訳例］

　航空機局が航空局と連絡設定できない場合には、航空機局は他の適切な周波数による連絡設定を試みること。この方法でも出来ない場合には、航空機局は、他の航空機または他の航空局との間の連絡設定を試みること。

(2) 一方送信 (Transmitting blind)

If the attempts specified in （1） above fail, the aircraft station shall transmit its message twice with the preceding phrase TRANSMITTING BLIND.

〔用語〕

preceding phrase 前置した語句

〔訳例〕

前項（1）に示した試みが出来なかった場合、航空機局は TRANSMITTING BLIND の語を前置し、その後に通報を2度送信すること。

(3) 受信機故障時の一方送信

(Transmitting blind due to receiver failure)

When an aircraft station is unable to establish communication due to receiver failure, it shall transmit message with the preceding phrase TRANSMITTING BLIND DUE TO RECEIVER FAILURE.

〔用語〕

receiver failure 受信機の故障

establish communication 通信を設定する

〔訳例〕

航空機局が受信機故障のために通信設定できない場合には、TRANSMITTING BLIND DUE TO RECEIVER FAILURE の語を前置して、通報を送信すること。

(4) SSR コードのセット (Selection of SSR code)

When an aircraft is unable to establish communication due to airborne equipment failure, it shall select the appropriate SSR code to indicate radio failure.

[用語]

airborne equipment failure 機上装備の故障　radio failure 無線機の故障

[訳例]

　航空機が機上装置の故障により通信設定できない場合には、通信機器の故障であることを知らせるために、該当する SSR コードをセットすること。

[5]　セルコールの運用（Operation of SELCAL）

(1)　セルコールの概略（An outline of SELCAL）

　SELCAL system replaces voice calling by an aeronautical station with transmission of coded tones to the aircraft. Tones are generated in the aeronautical station coder and are received by a decoder connected to the airborne receiver. Receipt of the assigned SELCAL code activates a call system in the cockpit.

[用語]

voice calling 音声による呼出し　coded tone コード化された音

aeronautical station coder 地上局のコーダー

assigned SELCAL code 割当てられたセルコールのコード

call system 呼出し装置

[訳例]

　SELCALシステムは、航空局からの音声による呼出しを、コード化された音の信号を航空機に対して送信することに置き換えるものである。音の信号は、航空局のコーダーによって発生され、機上受信機に取り付けられたデコーダーによって受信される。所定のSELCAL コードが受信されると、コックピットにある呼出しシステムが作動する。

（2） セルコールによる通信設定
（Communication establishment by SELCAL）

When an aeronautical station initiates a call by SELCAL, the aircraft station shall reply with the radio call sign followed by the phrase GO AHEAD.

［訳例］

　航空局がSELCALで呼出しを行った場合には、航空機局は、自局のコールサインとそれに続くGO AHEADの語で応答する。

[6]　遭難・緊急通信 （Distress and urgency communication）
（1）　遭難・緊急状態の定義
（Definition of distress and urgency conditions）

①　Distress：a condition of being threatened by serious and/or imminent danger and of requiring immediate assistance.

②　Urgency：a condition concerning the safety of an aircraft or other vehicles, or of some person on board or within sight, but which does not require immediate assistance.

［用語］

distress communication 遭難通信

urgency communication 緊急通信

serious danger 重大な危険　　imminent danger 急迫な危険

within sight 視界内　　immediate assistance 即座の支援

［訳例］

①　遭難：重大で急迫な危険にさらされており、かつ即時の援助を必要とする状態

②　緊急：航空機または他の乗り物あるいはそれらに搭乗ないしは視界内にいる人の安全に係わるが、しかし即時の援助を必要とはしない状態

(2) 遭難・緊急信号 (MAYDAY/PAN-PAN)

The aircraft station in distress or urgency shall transmit the distress signal MAYDAY or the urgency signal PAN-PAN respectively, preferably spoken three times, at the commencement of the first distress or urgency communication.

[用語]

distress signal 遭難信号　　distress communication 遭難通信

urgency signal 緊急信号　　urgency communication 緊急通信

[訳例]

遭難または緊急状態にある航空機局は、遭難信号MAYDAYまたは緊急信号PAN-PANを出来れば3回繰り返して、最初の遭難通信または緊急通信の開始時に送信すること。

(3) 通報に対する制限 (Restrictions on messages)

The originator of messages addressed to an aircraft in distress or urgency condition shall restrict to the minimum the number and volume and content of such messages as required by the condition.

[用語]

aircraft in distress condition 遭難状態にある航空機

aircraft in urgency condition 緊急状態にある航空機

[訳例]

遭難または緊急状態にある航空機に送信する通報は、数・量およびその内容がその時の状況に応じる最小限度に制限されるべきである。

(4) 使用周波数 (Frequencies to be used)

The aircraft station in distress or urgency shall initiate the distress and the urgency communication on the air-ground frequency in use and shall maintain it unless the station in control of such communication suggests frequency change to the emergency frequency 121.5MHz or other frequencies.

64

［用語］

aircraft station in distress 遭難状態にある航空機局

aircraft station in urgency 緊急状態にある航空機局

air-ground frequency in use 使用中の空地通信周波数

station in control 管轄している局

emergency frequency 非常用周波数

［訳例］

遭難または緊急状態にある航空機局は、遭難通信および緊急通信はその時使用中の空地通信周波数で開始し、遭難通信および緊急通信を管轄している局が、非常用周波数121.5MHzまたは他の周波数への変更を指示しない限り、その周波数を維持する。

(5) 送信要領 (Transmitting technique)

In case of distress and urgency communications, in general, transmissions by radiotelephony shall be made slowly and distinctly, each word being clearly pronounced to facilitate transcription.

［訳例］

遭難通信及び緊急通信の場合において、一般的に、無線電話による送信は、ゆっくり、かつ、明白に行い、各語は、筆記を容易にするためにはっきりと発音しなければならない。

[6]-1　遭難通信 (Distress communication)

(1)　遭難通報の内容 (Contents of distress message)

The distress message shall consist of as many as possible of the following elements, if possible, in the following order.

① 　Name of the station addressed (time and circumstances permissible)

② 　The identification of the aircraft

③ 　The nature of the distress condition

④　Intention of the pilot in command

⑤　Present position, level and heading

［用語］

distress message 遭難通報　　in the following order 次の順序で

［訳例］

　遭難通報には、下記の要素を可能な限り多く含み、もし可能ならば次の順序に従っていること。

①　送信相手局名（時間と状況の許す範囲で）

②　航空機のコールサイン

③　遭難状態の特徴

④　機長の意図するところ

⑤　現在位置、高度、および進行方向

(2)　SSR コードのセット (Selection of SSR code)

　Selection of appropriate SSR code gives the sign of the aircraft in distress condition.

［訳例］

　該当するSSRのコードをセットすることは、航空機が遭難状態にあることを示す合図を送信することになる。

(3)　遭難宛先局の措置 (Actions by the station addressed)

　The station addressed by aircraft in distress or the first station acknowledging the distress message shall :

①　Immediately acknowledge the distress message.

②　Take immediate action to ensure that all necessary information is made available to the ATS unit concerned and the aircraft operating agency. The requirement to inform aircraft operating agency does not have priority over other traffic control communications.

［用語］

station addressed 受信を指定された局

aircraft in distress 遭難状態の航空機

first station acknowledging the distress message　遭難通報を認めた最初の局

aircraft operating agency 航空機運航者

［訳例］

遭難状態の航空機から受信局と指定された局または最初に遭難通報を認めた局は、

① 直ちに受信を承認する。

② すべての必要な情報が関係する ATS 機関と航空機運航者とにより把握されていることを確保するための行動を直ちにとる。航空機運航者への知らせは、他の管制通信に対して優先するものではない。

(4)　沈黙の賦課（Imposition of silence）

The station in distress or the station in control of distress traffic is permitted to impose silence on all stations or on any station that interferes with distress traffic. It shall address STOP TRANSMITTING to all stations or a certain station.

［用語］

station in control of distress traffic 遭難通信を管轄している局

［訳例］

遭難状態の局または遭難通信を管轄している局は、すべての局または遭難通信の妨げとなる局に対して、沈黙を課することが許される。STOP TRANSMITTING をすべての局またはある局に対して発信する。

(5)　他の局による措置（Actions by the other stations）

The distress communications have absolute priority over all other communications. The station aware of them shall not transmit on the frequency concerned, and continue listening to such traffic until it is evident that assistance is being provided.

［用語］

absolute priority　絶対的優先度

station aware of them それらに気付いている局

frequency concerned　関係している周波数

［訳例］

　遭難通信には他の通信に対して絶対的優先度がある。遭難通信を知った局は、関係している周波数で送信してはならない、また遭難に対する援助が与えられていることが明白になるまでは、遭難通信を聴守しなければならない。

(6)　遭難の終了 (Termination of distress)

①　When an aircraft is no longer in distress, it shall transmit a message cancelling the distress condition.

②　When the station that has controlled the distress traffic becomes aware that the distress condition has ended, it shall immediately inform the stations concerned that the distress condition has ended.

③　The distress communication and silence condition shall be terminated by transmitting a message including the words DISTRESS TRAFFIC ENDED.

［訳例］

①　航空機は遭難から脱した場合には、遭難状態解消の通報を送信する。

②　遭難通信を管轄した局は遭難状態の終了を知った場合には、直ちに関係局に対して遭難状態が終了したことを知らせる。

③　遭難通信および沈黙状態は、DISTRESS TRAFFIC ENDED の語を含む通報を送信することによって終了される。

[6]-2　緊急通信 (Urgency communication)
（1）　緊急状態の航空機の取るべき措置
(Action by the aircraft in an urgency condition)

An aircraft in urgency condition shall transmit an urgency message containing the following elements as many as possible, in addition to the preceded urgency signal PAN-PAN preferably spoken three times, on the air-ground frequency in use at the time. The message shall be spoken distinctly and, if possible, in the following order.

① the name of the station addressed

② the identification of the aircraft

③ the nature of the urgency condition

④ the intention of the pilot in command

⑤ present position, level and heading

⑥ any other useful information

［訳例］

　　緊急状態の航空機は、前置きした緊急信号PAN-PAN （できるだけ3回） に加えて、可能な限り下記の要素を含む緊急通報を、その時使用中の空地通信周波数で送信する。通報は、はっきりと、またもし可能ならば、次の順序で送信されるものとする。

① 送信相手局名

② 航空機コールサイン

③ 緊急状態の特徴

④ 機長の意図

⑤ 現在位置、高度および進行方向

⑥ その他の有用な情報

（2）　緊急宛先局による措置 (Actions by the station addressed)

The station addressed by an aircraft reporting an urgency

condition or the first station acknowledging the urgency message shall：

① 　acknowledge the urgency message.

② 　take immediate action to ensure that all necessary information is made available to the ATS unit concerned, and the aircraft operating agency concerned. The requirement to inform the aircraft operating agency does not have priority over other safety communication.

③ 　if necessary, exercise control of communications.

［用語］

ATS unit concerned 関係する ATS 機関

control of communication 通信の管制

［訳例］

　緊急状態を通報している航空機により通信宛先局と指定された局、または緊急通報を最初に知った局は、以下のことを実施する。

① 　緊急通報の受信を確認する。

② 　必要なすべての情報が ATS 機関および航空機運航者に提供された状態にあることを直ちに確認する。航空機運航者に知らせることは他の安全通信に対して優先するものではない。

③ 　もし必要であれば、交信の管轄を行う。

(3)　他の局による措置（Actions by other stations）

　The urgency communications have priority over all other communications except distress conditions. All stations shall take care not to interfere with the transmission of urgency traffic.

［訳例］

　緊急通信は遭難を除くすべての通信に対して優先される。他のすべての局は、緊急通信に妨害となる送信をしないように注意しなければならない。

[7]　非合法的妨害時の通信
(Communication - unlawful interference)

The station addressed by an aircraft subjected to an act of unlawful interference, or the first station acknowledging a call from such aircraft, shall render all possible assistance.

〔用語〕

station addressed 通報を受けた局

an aircraft subjected to an act of unlawful interference 非合法的妨害行為に曝されている航空機

first station acknowledging a call 呼出しを最初に認識した局

all possible assistance あらゆる可能な支援

〔訳例〕

　非合法的妨害行為にさらされている航空機からの通報を受けた局、またはそのような航空機からの呼出しを最初に知った局は、あらゆる可能な支援を与えるべきである。

[8]　放送業務 (Broadcasting service)

Broadcasts shall be made on specified frequencies and at specified times. Schedules and frequencies of all broadcasts shall be publicized in appropriate documents. Any change in a frequency or time shall be publicized in NOTAM at least two weeks in advance of the change. Additionally, any such change shall, if possible, be announced on all regular broadcasts for 48 hours preceding the change and shall be transmitted once at the beginning and once at the end of each broadcast.

〔用語〕

broadcasting service 放送業務　　　specified frequency 指定周波数
specified time 指定時間　　　　　　regular broadcast 定例の放送
once at the beginning 開始時に 1 回　once at the end 終了時に 1 回

［訳例］

　放送は、指定周波数で指定時間に実施される。すべての放送のスケジュールと周波数は、適切な書類によって公表される。周波数および時間の変更は、少なくとも変更の2週間前までにNOTAMで公表されるべきである。また、そのような変更は、もし可能であれば、変更に先立つ48時間にわたってすべての定期放送において、各放送のはじめに1回と終りに1回公表されるものとする。

第４節のまとめ英文（The summary of section 4）

① Acknowledgement of receipt

An aircraft station shall acknowledge receipt of important ATC messages by read-back followed by its call sign. Flight clearance is subject to read back.

② End of conversation

A radio communication conversation is terminated by the receiving station using its call sign.

③ Correction

If a correction can best be made by repeating the entire message, CORRECTION, I SAY AGAIN shall be spoken before transmitting the correct message.

④ Emphasis

The transmitting person shall emphasize important element of the message by sending it twice.

⑤ Communication watch

An aircraft station shall maintain communication watch during flight except for reasons of safety.

⑥ Communication failure

If attempts on communication establishment fail, the aircraft station shall transmit a message twice with the words

TRANSMITTING BLIND, and select its SSR code.

⑦ Distress and urgency

An aircraft in distress or urgency shall transmit MAYDAY or PAN-PAN respectively, preferably spoken three times.

The aircraft in distress or urgency shall initiate calling on the air-ground frequency in use and maintain it unless the station controlling the communication suggests to change.

⑧ Broadcasting service

Broadcasts are made on specified frequencies and times. Any change to them shall be publicized in NOTAM at least two weeks in advance of the change.

┌─ 第4章 ─┐
航空交通管制
Air Traffic Control

　我が国の航空交通管制方式は、国際的に標準化された ICAO 方式であり、世界的運用に適用されるものである。この章では ICAO 資料を参照して管制、航法および監視について概説し、それらに係わる英語の習得に資することとする。

Japan's ATC procedures are internationally standardized, which is consistent with ICAO's provisions, and is applicable to worldwide operations. Brief explanation on air traffic control, navigation and surveillance are given in this chapter, intending promotion of English understanding related to such subjects.

第1節　管制 (ATC)

[1]　航空交通業務 (Air traffic service)

(1)　航空交通管制業務の目的 (Objectives of the air traffic services)

The objectives of the air traffic services shall be to :

① 　prevent collisions between aircraft

② 　prevent collisions between aircraft on the maneuvering area and obstructions on that area

③ 　expedite and maintain an orderly flow of air traffic

④ 　provide advice and information useful for the safe and efficient conduct of flights

⑤ 　notify appropriate organizations regarding aircraft in need of search and rescue aid, and assist such organizations as required

［用語］

air traffic service 航空交通業務　　maneuvering area 走行区域

74

air traffic 航空交通　　　　　　orderly flow 秩序ある流れ

［訳例］

① 航空機相互間の衝突防止

② 走行区域内における航空機と障害物との衝突防止

③ 航空交通の秩序ある流れを促進し維持すること

④ 安全かつ効率的な運航に有用な助言と情報を提供すること

⑤ 捜索救難援助が必要な航空機に関して適当な機関に通報し、必要に応じて当該機関を支援すること

(2) 航空交通業務の分類 (Classification of air traffic service)

The air traffic services are classified into air traffic control service, flight information service and alerting service.

① Air traffic control service promotes safe and efficient flights providing the service for IFR aircraft operating in the control area and the control zone, and VFR flights as well.

② Flight information service provides aircraft with weather information, operational conditions of navigation facilities, conditions of airports and their facilities, and traffic information.

③ Alerting service is initiated by the unit providing air traffic service when it becomes aware of an aircraft in emergency condition.

［用語］

air traffic control service 航空交通管制業務

flight information service 飛行情報業務　　alerting service 警急業務

control area 航空交通管制区　　　　　control zone 航空交通管制圏

operational condition 運用状態　　　　navigation facility 航法施設

traffic information 交通情報　　　　　emergency condition 非常状態

［訳例］

　航空交通業務は、航空交通管制業務、飛行情報業務および警急業務に分類される。

①　航空交通管制業務は、航空機の安全かつ効率的な運航を促進し、その業務は管制空域および管制圏を航行する IFR 航空機ならびに VFR 航空機にも提供される。

②　飛行情報業務は、航空機に対して気象情報、航法施設の運用状態、飛行場とその施設の状態ならびに交通に関する情報を提供する。

③　警急業務は、ある航空機が非常状態になったことを知った航空交通業務を提供している機関が開始する。

(3)　航空交通業務の提供 (Provision of air traffic service)

Air traffic services are provided for the flights operating in the controlled airspace and Fukuoka flight information region (FIR).

［用語］

controlled airspace 管制空域

flight information region（FIR）飛行情報区

［訳例］

航空交通業務は、管制空域および福岡情報区内の運航に対して提供される。

［2］　管制空域・飛行情報区 (Controlled airspace and FIR)

The controlled airspace consists of the control area that includes the approach control area, the positive control area (PCA) and the terminal control area (TCA), the control zone and the oceanic control area of Fukuoka FIR.

(1)　航空交通管制区・進入管制区

(Control area and approach control area)

①　航空交通管制区 (Control area)

All the areas inside the QNH setting line from the following lower limits to the unlimited height.

a)　Upper limit of the control zone when interfaced with the control zone.

b) 700ft above the surface within 20nm radius of the aerodrome reference point of the airport wherein any instrument approach procedure or standard instrument departure is published.

c) 1,000ft above the surface within 40nm radius of the aerodrome reference point of the airport except b) wherein the approach control area is published.

d) 2,000ft above the surface when neither b) nor c) are applicable.

② 進入管制区 (Approach control area)

A part of the control area congested with IFR departure and arrival aircraft, is published as the approach control area. The terminal radar service is provided within this area.

③ ターミナルコントロールエリア (Terminal control area)

Among the approach control airspace, the areas congested with VFR are published as the Terminal Control Area (TCA) wherein TCA advisory service is provided for VFR aircraft.

(2) 航空交通管制圏 (Control zone)

Control zone is published by the MLIT in order to ensure safety of takeoff and landing aircraft. A control zone extends 5nm radius of the aerodrome reference point, and above the surface to the specified altitude (MSL). Generally, upper limit is 3,000ft at civil airport.

(3) 航空交通情報圏 (Information zone)

An information zone is the airspace designated by MLIT in order to secure the safety of arrival and departure aircraft in and out of the airport within the zone. The zone is principally specified by the relevant Public Notice for an airport without having the control zone but IFR operations are conducted.

(4) 特別管制区 (Positive control area)

An aircraft is required to operate under IFR within the Positive controlled airspace published as Positive Control Areas unless otherwise authorized by ATC.

(5) 洋上管制区 (Oceanic control area)

The oceanic control area generally begins from 5,500ft above the oceanic surface out of QNH airspace within Fukuoka FIR.

［用語］

approach control area 進入管制区

terminal control area ターミナルコントロールエリア

oceanic control area 洋上管制区　upper limit 上限　surface 地表面

aerodrome reference point 飛行場標点

MLIT (Ministry of Land, Infrastructure, Transport and Tourism)　国土交通省

MSL (mean sea level) 平均海面高度

relevant Public Notice 関連する公示　　oceanic surface 海面

［訳例］

　管制空域は、進入管制区、ターミナルコントロールエリア、特別管制区を含む航空交通管制区、航空交通管制圏および福岡FIRの洋上管制区から成る。

(1) 航空交通管制区・進入管制区

① 航空交通管制区とは、QNH適用区域境界線の内側全ての区域で、上限はなく下限の高さを次のいずれかとする空域をいう。

a) 管制圏の上空については、当該管制圏の上限高度

b) 計器飛行による進入方式または出発方式が設定されている飛行場の標点を中心とする半径20マイルの円内の区域については地表から700ft

 c) 進入管制区が指定されている飛行場において、b) の区域の外側で、飛行場の標点を中心とする半径 40 マイルの円内の区域については地表から 1,000ft

 d) 上記 b) にも上記 c) にも該当しない区域については地表から 2,000ft

② 管制区の中で、計器飛行方式による出発機および到着機の多い区域は進入管制区として告示されており、この空域ではターミナル・レーダー管制業務が提供されている。

③ 進入管制区のうち、特に VFR 機の輻輳する空域では、VFR 機に対して TCA アドバイザリー業務を実施する空域がターミナルコントロールエリアとして公示されている。

(2) 航空交通管制圏

 航空交通管制圏とは、管制圏内の飛行場に離発着する航空機の航行の安全のために国土交通大臣が告示で指定した空域である。管制圏は飛行場の標点から半径 5 マイルの円で囲まれる区域の上空で、民間の飛行場にあっては通常 3,000ft MSL（平均海面高度）である。

(3) 航空交通情報圏

 航空交通情報圏とは、情報圏内の空港に離発着する航空機の航行の安全のために国土交通大臣が告示で指定した空域をいう。原則として管制圏が指定されていない空港のうち、IFR による離発着が行える飛行場に関連する公示により指定されている。

(4) 特別管制区

 特別管制空域は、管制機関から許可された場合を除き IFR で飛行しなければならない空域であり、個々の空域は特別管制区（Positive control area）として公示されている。

(5) 洋上管制区

 洋上管制区とは、福岡 FIR の洋上区域であって、QNH 適用区域境界線の外側にあり、原則として海面から 5,500ft 以上の空間をい

う。

[3]　航空交通管制業務（Air traffic control service）
(1)　航空交通管制業務（Air traffic control service）

Air traffic control services are divided into five categories in accordance with the nature of service.

① En-route air traffic control service : This service is provided for aircraft operating on IFR flight plans and aircraft within the positive control area, excluding ②, ③, ④ and ⑤.

② Aerodrome control service : This service is provided for aircraft arriving to and departing from an airport, aircraft flying in the airport vicinity and personnel or vehicles on the airport surface, excluding ③, ④ and ⑤.

③ Approach control service : This service is provided for arriving and departing IFR aircraft or any aircraft operating within PCA in order to ensure the IFR separation among IFR aircraft excluding ④ and ⑤.

④ Terminal radar control service : This service is provided for aircraft mentioned in ③ utilizing radar equipment excluding ⑤.

⑤ Ground controlled approach service : This service is provided for IFR arriving aircraft applying azimuth and elevation guidance information obtained by radar equipment for precision approach.

［用語］

en-route air traffic control service　航空路管制業務

aerodrome control 飛行場管制　　approach control 進入管制

ground controlled approach service 着陸誘導管制業務

azimuth 方位　　elevation 高度

guidance 誘導　　precision approach 精密進入

［訳例］

航空交通管制業務はその業務の分担範囲によって5つに分類される。

① 航空路管制業務：この業務は、IFR により飛行する航空機および特別管制空域を飛行する航空機に対する管制業務であって、後述する②,③,④,⑤を除く業務をいう。

② 飛行場管制業務：この業務は、飛行場において到着および出発機、飛行場周辺を飛行する航空機、または飛行場の業務に従事する者および車両に対する管制業務であって、後述する③,④,⑤を除く業務をいう。

③ 進入管制業務：この業務は、到着および出発 IFR 機または特別管制空域を飛行する航空機に対して、IFR の安全間隔を保証しつつ、管制業務を提供する。後述する④,⑤を除く。

④ ターミナルレーダー管制業務：③の航空機に対してレーダーを使用して行う管制業務であって、次の⑤を除く業務をいう。

⑤ 着陸誘導管制業務：この業務は、精密進入のためのレーダーにより得られた方位および高度の誘導情報を適用して、到着 IFR 機に対して行う管制業務をいう。

（2）　航空交通管制機関（ATC facility）

Air traffic control facility are classified into five varieties in accordance with the nature of service.

① Air Traffic Management Center

　　・air traffic management service

② Area Control Center

　　・en-route air traffic control service

　　・approach control service

③ Terminal Control Facility

　　・approach control service

　　・terminal radar control service

④　Aerodrome Control Facility

　　・aerodrome control service

⑤　Ground Controlled Approach Facility

　　・ground controlled approach service

All of these facilities provide the flight information and alerting services in addition to their own ATC services. conducted.

[用語]

ATC facility　航空交通管制機関

Air Traffic Management Center　航空交通管理センター

Area Control Center　管制区管制所

en-route air traffic control service　航空路管制業務

approach control service　進入管制業務

Terminal Control Facility　ターミナル管制所

terminal radar control service　ターミナルレーダー管制業務

Aerodrome Control Facility　飛行場管制所

Ground Controlled Approach Facility　着陸誘導管制所

alerting service　警急業務

[訳例]

　　管制業務を実施する機関を管制所という。管制所はその業務内容によって次の5つに分類される。

①　航空交通管管理センター

　　・航空交通管管理管制業務

②　管制区管制所

　　・航空路管制業務

　　・進入管制業務

③　ターミナル管制所

　　・進入管制業務

　　・ターミナルレーダー管制業務

④　飛行場管制所

　　・飛行場管制業務

⑤　着陸誘導管制所

　　・着陸誘導管制業務

　各管制所ともそれぞれの管制業務を行うほか、あわせて飛行情報業務および警急業務をも行う。

[4]　情報の提供 (Provision of information)

(1)　気象情報 (Weather information)

①　ATS units provide pilots with weather information related to the wind data, visibility and runway visual range (RVR) with general weather conditions observed by a tower, a pilot in flight, and radar, in addition to the information issued by the weather agency.

②　The words "tower observation" or "pilot report" is prefixed to information of tower observation or pilot's report respectively.

③　Significant weather conditions such as moderate to severe turbulence, icing, active thunderstorm and low level windshear are included in weather information.

〔用語〕

weather information 気象情報　　　wind data 風のデータ

visibility 視程　　　　　　　general weather condition 全般的な気象状態

weather agency 気象機関　　　turbulence 擾乱　　　　icing 着氷

low level windshear 低高度におけるウィンドシア

runway visual range (RVR) 滑走路視距離

〔訳例〕

①　ATS の機関は、パイロットに、風のデータ、視程、滑走路視距離に関する気象情報とタワー観測、飛行中のパイロットの観測、レーダー観測によって得られた一般的気象状態に気象機関の発行する情報を加えて提供する。

② タワーの観測およびパイロットの報告には、それぞれ「タワー観測」または「パイロット報告」を示す用語が前置きされる。

③ 中程度から激しい程度の擾乱、着氷、活発な雷を伴った嵐および低層ウィンドシアのような顕著な気象状態は、気象情報の中に含まれる。

(2) 機上報告 (Weather report by pilots (PIREP))

A pilot should report weather conditions such as CAT, turbulence, thunderstorm, icing, windshear and other conditions that may adversely affect flight operations.

[用語]

CAT 晴天乱気流

[訳例]

パイロットは、CAT、擾乱、雷を伴う嵐、着氷、ウィンドシアのような気象状態およびその他の運航に悪影響を与えると思われる気象状態は、報告しなければならない。

(3) 飛行場情報放送業務 (ATIS)

ATIS is a broadcast of terminal information that contains type of approach, runway in use, weather information and operational conditions of navigation facilities. It is periodically updated during regular operation, and frequently updated during rapidly changing weather conditions. ATIS is available on VHF broadcasting system and VHF data link as well.

[用語]

ATIS 飛行場情報放送業務	terminal information 飛行場情報
type of approach 進入方式	runway in use 使用滑走路
operational condition 運用状態	regular operation 通常の運用

[訳例]

ATISは、進入方式、使用中の滑走路、気象情報および航法施設の運用状態を含む飛行場情報放送である。通常の運用では定時的にアップデートされ、急速に変化しつつある気象状態では頻繁にアップデート

される。ATISは、VHFの放送およびVHFのデータリンクにより提供されている。

第2節：航法と監視（Navigation and Surveillance）

（1）CNS/ATM

CNS/ATM application to flight operations is progressing, but conventional navigation and surveillance system generally controls air traffic of domestic operations. CNS/ATM system employs satellite navigation and communication. The conventional system is based on VHF/HF voice communication, VHF/UHF radio navigation, and radar surveillance. The following table indicates comparison of CNS/ATM and the conventional system.

	Conventional system		CNS/ATM	
	Within radar coverage	Beyond radar coverage	Within radar coverage	Beyond radar coverage
Communication	VHF voice	HF voice	VHF/SSR VHF voice	CPDLC INMARSAT
Navigation	VOR/DME	IRS	GPS, IRS	GPS, IRS
Surveillance	radar	HF voice position report	radar	ADS

〔用語〕

flight operations 運航

navigation and surveillance system 航法・監視システム

domestic operations 国内運航

satellite navigation and communication 衛星利用の航法・通信

voice communication 音声通信　　　radio navigation 無線航法

radar surveillance　レーダー監視

conventional system　従来のシステム

radar coverage　レーダーカバー範囲

［訳例］

　　CNS/ATMシステムの運航に対する適用は進んでいる。しかしながら
国内航空交通では従来の航法・監視システムが全般的に行われている。
CNS/ATMシステムは、衛星利用の航法・通信を採用している。従来シ
ステムは、VHF/HFの音声通信、VHF/UHFの無線航法およびレーダー
による監視である。次の表は、CNS/ATMシステムと従来システムの比
較を示す。

	従来システム		CNS/ATM	
	レーダーカバー範囲内	レーダーカバー以遠	レーダーカバー範囲内	レーダーカバー以遠
通信	VHF 音声	HF 音声	VHF/SSR VHF 音声	CPDLC INMARSAT
航法	VOR/DME	IRS	GPS, IRS	GPS, IRS
監視	レーダーによる監視	HF 音声による位置通報	レーダーによる監視	ADS

(2)　通信 (Communication)

　　Conventional communication system for ATC services consists
of VHF voice system for domestic flight operations, and HF voice
system for long range flight operations. Communication in CNS/ATM
system includes satellite communication system besides the
conventional voice communication system.

［用語］

long range flight operation　長距離の運航

［訳例］

　ATC業務の従来型の通信は、国内航空のためのVHF音声通信および長距離運航のためのHF音声通信から成る。CNS/ATMシステムにおける通信は、従来の音声通信のほかに衛星通信システムをも含む。

（3）　航法（Navigation）

RNAV route system permits pilots to fly any optional route without imposing restriction of ground based navaids. The system conducts high quality navigation with combined navigation information of the ground based facilities and airborne navigation equipment （IRS） or global navigation satellite system.

［用語］

optional route 随意の航路

ground based navaids 地上装備の航法支援施設

airborne navigation equipment 機上航法装置

global navigation satellite system 衛星航法システム

［訳例］

　RNAVルートシステムは、パイロットが地上装備の航法支援システムによる制約を受けることなく、所望の随意の航路を飛ぶことを可能とする。このシステムは、地上施設の航法情報と機上の航法装備（IRS）または衛星航法システムの航法情報を兼ね合わせた航法情報により、品質の高い航法を実施する。

（4）　監視（Surveillance）

Conventional surveillance is conducted by use of radar systems such as ASR, ARSR and ORSR of which coverage is limited. Surveillance beyond radar coverage is impossible. Automatic dependent surveillance （ADS） has become available by use of digital data position reporting system.

［用語］

conventional surveillance 従来型の監視

radar coverage レーダー覆域

automatic dependent surveillance 自動従属監視

position reporting　位置通報

［訳例］

　従来型の監視は、ASR、ARSRあるいはORSRといった覆域に制限の
あるレーダーによって実施される。覆域以遠の監視は出来ない。自動
従属監視は、デジタルデーター位置通報システムを利用することに
よって可能になった。

第4章のまとめ英文（The summary of chapter 4）

① The essential object of the air traffic service is to prevent collisions of aircraft.

② Air traffic services activity consists of air traffic control services,
flight information services and alerting services.

③ Air traffic services are provided for the aircraft operating in the
controlled airspace and Fukuoka FIR.

④ The controlled airspace is divided mainly into the control area
that includes approach control area, and the control zone.

⑤ ATC units provide ATC services and alerting service.

⑥ ACC provides en-route service and approach control service.

⑦ Aerodrome control unit, the tower is divided into delivery section, ground control section and the tower at major airports.

⑧ The delivery section relays the flight clearance to the aircraft in
preparation for departure.

⑨ The tower provides the aircraft arriving and departing and operating in the control zone with traffic control services.

⑩ The ATIS is a broadcasting service of terminal information that
includes type of approach, runway in use, weather information and
operational conditions of navigation facilities.

第5章
航空無線通信の実務
Radio Communication Practice

　IFR フライトは、飛行中 ATC のクリアランスおよびその指示に従って運航される。したがってパイロットと管制官との間に通信が維持されなければならない。この章では、飛行の各段階における通信手順を概説し、ATC 英語の習得に資することとする。

　IFR flight operation requires communication between pilots and controllers because it is operated in compliance with ATC clearance and instruction throughout the flight. This chapter intends to promote understanding of ATC English by presenting an outline of communication procedures of each flight phase.

第 1 節：出発前 (Pre-Departure)

(1) 飛行計画のファイル (Filing the flight plan)

　A pilot should file a flight plan before starting flight. In case of IFR operation, the flight plan should be filed at least 30 minutes prior to ETD. Filing the flight plan for VFR operation within the area 9km radius of the departing point is exempted. The flight plan is reviewed by ACC and issued as a flight clearance.

［用語］

flight plan 飛行計画　ETD 予定出発時刻　departing point 出発地点

IFR operation IFR による運航　VFR operation VFR による運航

［訳例］

　　パイロットは、飛行に先立ち飛行計画をファイルしなければならな

い。IFR運航の場合には、飛行計画は少なくともETDの30分前までにファイルされなければならない。出発地点の半径9km以内のVFR飛行に対する飛行計画のファイルは、免除される。飛行計画は、ACCによって検討され、フライトクリアランスとして発行される。

(2) ATC クリアランスの伝達 (ATC clearance delivery)

A pilot of an IFR operation will request an ATC clearance to the delivery section of a tower at around 5 minutes before ETD stating destination, proposed altitude and the parking location. The delivery section will relay the ATC clearance to the pilot. The pilot should read back the contents of the clearance.

　　Pilot：Tokyo Delivery, Spaceair 101, Osaka, flight level 240, spot 3.

　　ATC：Spaceair 101, Tokyo delivery, Clearance (contents of the clearance) read back.

〔用語〕

delivery section 管制承認伝達席　　proposed altitude 所望の高度

parking location 駐機場所　　　　　read back 復唱する

〔訳例〕

　IFR運航のパイロットは、ETDの約5分前にタワーの管制承認伝達席に対して、目的地、予定高度および駐機場所を述べてATCクリアランスを要求する。管制承認伝達席は、ATCクリアランスをパイロットに伝達する。パイロットは、クリアランスの内容を復唱しなければならない。

(3) タクシー (Taxi)

① A pilot will request ATC taxi instruction, stating his/her position with ATIS code.

　　Pilot：Tokyo Ground, Spaceair 101, spot 3, request taxi information Alfa.

② Taxi instruction includes the runway in use, wind, QNH, and other information.

Ground control: Spaceair 101, runway 34 right, wind 290 at 8,
QNH 2986, taxi via hotel 1, golf, echo to holding
point runway 34 right.

③　The pilot should read back the runway number and taxi route.

④　Taxi limit：the pilot should stop before the stop line proximate
to the runway.

〔用語〕

taxi instruction タクシーの指示　　ATIS code　ATIS のコード
ground control 地上管制席　　　　　taxi limit タクシーの限界
stop line 停止線

〔訳例〕

①　パイロットは、ATC に対して現在位置と ATIS コードを述べてタ
クシーの指示を要求する。

Pilot：Tokyo Ground, Spaceair 101, spot 3, request taxi information Alfa.

②　タクシーの指示には、使用中の滑走路ナンバー、風、QNH その他
の情報が含まれる。

Ground control: Spaceair 101, runway 34 right, wind 290 at 8,
QNH 2986, taxi via hotel 1, golf, echo to holding
point runway 34 right.

③　パイロットは、滑走路ナンバーおよびタクシールートを復唱しなけ
ればならない。

④　タクシーの限界：パイロットは、滑走路に最も近い停止線の手前で
止まること。

(4) 出発手順 (Departure procedures)

①　Ground control will advise the pilot to contact tower.

Ground control：Spaceair 101, contact tower 118.1.

② The pilot contacts the tower.

Pilot：Tower, Spaceair 101, ready.

③ Tower gives the pilot the runway number and the instructions to proceed into the runway.

Tower：Spaceair 101, runway 34 right, line-up and wait.

〔訳例〕

① グラウンドコントロールは、パイロットにタワーにコンタクトするようにアドバイズする。

Ground control：Spaceair 101, contact tower 118.1.

② パイロットは、タワーにコンタクトする。

Pilot：Tower, Spaceair 101, ready.

③ タワーは、パイロットに、滑走路ナンバーを告げて滑走路へと進む指示を与える。

Tower：Spaceair 101, runway 34 right, line-up and wait.

(5) 離陸のクリアランス (Takeoff clearance)

① The tower issues takeoff clearance with wind information and the runway number.

Tower：Spaceair 101, wind 240 at 15, runway 34 right, cleared for takeoff.

② The pilot should read back the runway number and takeoff clearance.

Pilot：Spaceair 101, runway 34 right, cleared for takeoff.

〔用語〕

takeoff clearance 離陸のクリアランス

〔訳例〕

① タワーは、パイロットに風の情報と滑走路ナンバーを告げて離陸のクリアランスを与える。

Tower：Spaceair 101, wind 240 at 15, runway 34 right, cleared for takeoff.

② パイロットは、滑走路ナンバーおよび離陸のクリアランスを復唱しなければならない。

Pilot：Spaceair 101, runway 34 right, cleared for takeoff.

(6) 最低気象条件 (Weather minima)

Tower will not issue takeoff clearance when the weather minima are not met.

Tower：Spaceair 101, unable to issue departure clearance, RVR touchdown 100 meters, midpoint 400 meters and stop end 300 meters.

〔用語〕

weather minima 最低気象条件

departure clearance 出発のクリアランス

touchdown 接地点　midpoint 中間点　stop end 最終点

〔訳例〕

タワーは、最低気象条件が満たされていない時には離陸クリアランスは発行しない。

Tower：Spaceair 101, unable to issue departure clearance, RVR touchdown 100 meters, midpoint 400 meters and stop end 300 meters.

(7) 離陸の中止 (Cancellation of takeoff)

① A pilot will reject takeoff when necessary. He/she should immediately inform the tower of rejection and his/her intention with the reason of rejection.

Pilot：Spaceair 101 rejected takeoff, request taxi back to the parking spot, a fire warning was activated.

② Tower will cancel takeoff clearance by ordering the pilot to discontinue takeoff roll or to hold its position when imminent danger exists.

> Tower：Spaceair 101, stop immediately, Spaceair 101, stop immediately, traffic landing.

> Tower：Spaceair 101, hold position, cancel takeoff clearance, arrival traffic is going around.

③　Controllers refrain from calling pilots while they are taking off.

[用語]

reject takeoff 離陸の中断　　　　parking spot 駐機場

fire warning 火災警報　　　　　takeoff roll 離陸滑走

imminent danger 急迫な危険　　　arrival traffic 到着機

go-around 着陸復行

[訳例]

①　パイロットは、必要が生じた場合には離陸を中断する。パイロットは、離陸の中断と意図するところおよび離陸中断の理由を直ちにタワーに知らせる。

> Pilot：Spaceair 101 rejected takeoff, request taxi back to the parking spot, a fire warning was activated.

②　タワーは、急迫な危険がある時は、パイロットに離陸滑走を中断させあるいはその位置に止めさせて離陸クリアランスをキャンセルする。

> Tower：Spaceair 101, stop immediately, Spaceair 101, stop immediately, traffic landing.

> Tower：Spaceair 101, hold position, cancel takeoff clearance, arrival traffic is going around.

③　管制官は、パイロットが離陸中には通信呼出しはしない。

(8) 騒音軽減方式 (Noise abatement procedures)

Pilots are requested to comply with noise abatement procedures. The procedures are ① preferential runway procedure, ② steep climb procedure, ③ cutback climb procedure and ④ preferential departure route procedure.

〔用語〕

noise abatement procedure 騒音軽減方式

preferential runway procedure 優先滑走路方式

steep climb procedure 急上昇方式

cutback climb procedure カットバック上昇方式

preferential departure route procedure 優先出発経路方式

〔訳例〕

　パイロットは、騒音軽減方式に従うように要求されている。その方式は、①優先滑走路方式 ②急上昇方式 ③カットバック上昇方式および ④優先出発経路方式である。

第 1 節のまとめ英文 (The summary of section 1)

① A pilot should file a flight plan before starting flight. In case of IFR operation, the flight plan should be filed at least 30 minutes before ETD.

② A pilot of an IFR operation will request Delivery the ATC clearance at around 5 minutes before ETD. Delivery will relay ATC clearance to the pilot.

③ Taxi instruction includes the runway in use, wind, and QNH. Taxi limit is the hold line proximate to the runway.

④ A pilot will reject takeoff when necessary. He/she will immediately inform the tower of rejection and his/her intention with the reason of rejection.

⑤ A pilot should comply with noise abatement procedures. The procedures are (1) preferential runway, (2) steep climb, (3) cutback climb and (4) preferential route.

第2節：出発 (Departure)

(1) 離陸後の飛行手順 (Flight procedures after airborne)

① 標準出発経路と転移経路 (SID and transition)

ATC clearance usually includes SID and transition. Pilots will follow the procedures specified by them after takeoff until reaching the assigned route.

〔用語〕

flight procedure 飛行方式　　airborne 離陸後

assigned route 所定の航空路

〔訳例〕

ATCクリアランスには通常標準出発経路（SID）および転移経路が含まれる。パイロットは、離陸後所定の航空路に達するまではそれらに指定されている飛行手順に従う。

② レーダー誘導 (Radar vectoring)

Radar vectoring is initiated when radar identification is established.

Radar：Spaceair 101, Radar contact, fly present heading.

Pilot：Fly present heading, Spaceair 101.

〔用語〕

radar vectoring レーダー誘導　radar identification レーダーによる識別

radar contact レーダーによる把握　present heading 現在の機首方位

〔訳例〕

レーダー誘導は、レーダーによる識別が確立された時に開始される。

Radar：Spaceair 101, Radar contact, fly present heading.

Pilot：Fly present heading, Spaceair 101.

③ 出域管制との通信設定 (Contact departure control)

Pilots will contact departure control on the given frequency upon instruction by the tower.

Tower：Spaceair 101, contact Departure 120.8

Pilot：120.8, Spaceair 101.

Pilot：Tokyo departure, Spaceair 101, leaving 1,000 feet for FL150.

〔用語〕

departure control 出域管制

〔訳例〕

　パイロットは、タワーからの指示に従って所定の周波数で出域管制との通信設定を行う。

Tower：Spaceair 101, contact Departure 120.8

Pilot：120.8, Spaceair 101.

Pilot：Tokyo departure, Spaceair 101, leaving 1,000 feet for FL150.

④　レーダートラフィック情報の提供

（Provision of radar traffic information）

ATC may provide aircraft with radar traffic information.

ATC：Spaceair 101, traffic two o'clock, 10 miles, eastbound, Boeing 737, 12,000 feet.

Pilot：Looking out.

Pilot：Traffic in sight（or negative contact）.

ATC：Clear of traffic.

〔用語〕

radar traffic information レーダートラフィック情報

eastbound 東行き　　sight 視界　　negative contact 発見できない

〔訳例〕

　ATCは、航空機にレーダートラフィック情報を提供する。

ATC：Spaceair 101, traffic two o'clock, 10 miles, eastbound, Boeing 737, 12,000 feet.

Pilot：Looking out.

Pilot：Traffic in sight（or negative contact）.

ATC：Clear of traffic.

⑤　レーダー誘導の終了（Termination of radar vectoring）

Radar vectoring is terminated when it is no longer required.

ATC：Spaceair 101, resume own navigation, direct Sekiyado.

Pilot：Direct Sekiyado. Spaceair 101.

〔用語〕

resume 再び開始する

〔訳例〕

レーダー誘導は、その必要が無くなったら終了される。

ATC：Spaceair 101, resume own navigation, direct Sekiyado.

Pilot：Direct Sekiyado. Spaceair 101.

⑥　高度計のセット（Altimeter setting）

Pilots will change the altimeter setting from QNH to QNE when passing 14,000 feet during climb. QNE is based on the ISA （International Standard Atmosphere）.

〔訳例〕

パイロットは、上昇中14,000ftを通過する時に高度計のセットをQNHからQNEに変更する。QNEは、世界標準大気状態を基準としている。

⑦　レーダーサービスの移管（Transfer of radar service）

Radar service is continuously provided by transferring it from Departure to ACC, between ACC segments, and from ACC to Approach control.

〔用語〕

transfer 移管する　　continuously 継続的に

［訳例］

　レーダーサービスは、出域管制からACCへ、ACC区間相互間で、およびACCから進入管制へと移管することによって、継続的に提供される。

第2節のまとめ英文 (The summary of section 2)

① Radar vectoring is initiated when radar identification is established. A pilot should comply with vectoring instruction. Radar vectoring is terminated when it is no longer required.

② Pilots will change altimeter setting from QNH to QNE when climbing through 14,000ft. QNE is based on the international standard atmospheric condition (ISA).

第3節：エンルートの運航 (En-route Operations)

[1]　航路上のコース (En-route courses)

The en-route course consists of the transition routes, airways and RNAV routes.

(1) 転移経路 (Transition route)

The transition route is the route between the last fix of SID and the intercepting point with the airway.

(2) 航空路 (Airway)

The airway is a route connecting radio navigation stations such as VOR or NDB.

(3) RNAVルート (RNAV route)

RNAV routes are established in a radar service available area.

［用語］

enroute course 航路上のコース　　　　transition route 転移経路

the last fix of SID　SID の最終地点　　intercepting point 接点

radio navigation station 無線航法の局

［訳例］

航路上のコースは、転移経路、航空路およびRNAVルートから成る。

（1）　転移経路は、SID の最終フィックスから航空路との接点までの間の経路である。

（2）　航空路は、VOR や NDB のような航法無線局を連結したルートである。

（3）　RNAV ルートは、レーダーサービスが利用可能なところにおいて設定される。

［2］　航路上の運航（En-route operations）

（1）速度の調整（Speed adjustment）

①　Aircraft speed is expressed in IAS.

②　The following phrases will be used in speed adjustment:

ATC：Maintain 280 knots.

ATC：Maintain Mach point 82.

ATC：Increase（or reduce）speed to 300（or 240）knots.

ATC：Increase（or reduce）speed by 20 knots.

Pilot：Unable to maintain 320 knots due to turbulence, Spaceair 101.

［訳例］

①　航空機の速度は、IAS で表現される。

②　速度調整には、下記の語句が使用される。

ATC：Maintain 280 knots.

ATC：Maintain Mach point 82.

ATC：Increase（or decrease）speed to 300（or 240）knots.

ATC：Increase（or decrease）speed by 20 knots.

Pilot：Unable to maintain 320 knots due to turbulence, Spaceair 101.

(2) 通信移管 (Communication transfer)

① Radio frequencies are changed along with control transfer such as from Departure to ACC or between ACC sectors.

ATC：Spaceair 101, contact Tokyo control 124.1.

ATC：Spaceair 101, contact Naha control 132.3 at SABAN.

〔用語〕

radio frequency 無線周波数　　along with ~ に伴って

control transfer 管制移管

〔訳例〕

無線周波数は、出域管制からACCへあるいはACCセクション相互間などにおける管制移管に伴って変更される。

ATC：Spaceair 101, contact Tokyo control 124.1.

ATC：Spaceair 101, contact Naha control 132.3 at SABAN.

② Pilot will report the assigned flight level or altitude with present altitude on initial contact.

Pilot：Fukuoka control, Spaceair 101, flight level 370.

Pilot：Naha control, Spaceair 101, leaving 4,300 climbing, assigned 7,000.

〔用語〕

assigned flight level 指定高度　　present altitude 現在の高度

initial contact 最初の連絡設定

〔訳例〕

パイロットは、最初のコンタクト時に指定されたフライトレベルまたは高度を現在高度とともに報告する。

Pilot：Fukuoka control, Spaceair 101, flight level 370.

Pilot：Naha control, Spaceair 101, leaving 4,300 climbing,

assigned 7,000.

(3) インターパイロット通信 (Interpilot communication)

① Operational safety communication between aircraft stations may be conducted on the specified frequency.

② The term "interpilot" should be affixed to the call.

③ The company radio frequency should be used for interpilot communication between aircraft stations of the same airline.

　Pilot：Marsair 202, Spaceair 101, interpilot, do you read.

〔用語〕

operational safety communication 運航安全通信

specified frequency 所定の周波数　　　　company radio 会社用無線

〔訳例〕

① 航空機局間における運航の安全に関する通信は、所定の周波数を使って実施される。

② 呼出しに先立って interpilot の用語をつける。

③ 同じ航空会社の航空機相互間であれば、会社の周波数を用いること。

　Pilot：Marsair 202, Spaceair 101, interpilot, do you read.

(4) コースの逸脱 (Course deviation)

A pilot will request ATC course deviation when necessary for a case such as to avoid weather.

　Pilot：Request weather deviation to the right/left of track up to 10 miles due to Cb.

　Pilot：Request heading 340 to avoid weather.

〔用語〕

course deviation コースからの逸脱　　heading 機首方位

〔訳例〕

　悪天候を避けるなど必要な時にはパイロットは、ATCにコースからの逸脱を要求する。

Pilot：Request weather deviation to the right/left of track up to 10 miles due to Cb.

Pilot：Request heading 340 to avoid weather.

(5) 位置通報 (Position report)

① A pilot is requested to make a position report at the specified point.

Pilot：Tokyo Control, Spaceair 101, Okayama VOR, at 35, FL 370, estimating Kuga 41.

② PIREP：A pilot is requested to make PIREP when he/she has observed or encountered significant weather condition.（domestic flights）

Pilot：Spaceair 101, observed Cb, 20 miles northwest of Yamagata.

Pilot：Spaceair 101, encountered moderate CAT over Yamagata, FL 350, Boeing 737.

〔用語〕

position report 位置通報　　specified point 所定の地点

significant weather 悪天候

〔訳例〕

① パイロットは、所定の場所で位置通報をすることを要求されている。

Pilot：Tokyo Control, Spaceair 101, Okayama VOR, at 35, FL 370, estimating Kuga 41.

② PIREP：パイロットは、顕著な気象状態を観察またはそれに遭遇した場合には、PIREP により報告することを要求されている（国内運航）。

Pilot：Spaceair 101, observed Cb, 20 miles northwest of Yamagata.

Pilot：Spaceair 101, encountered moderate CAT over Yamagata, FL 350, Boeing 737.

(6) 降下のクリアランス (Descent clearance)

ATC will issue various types of descent clearances for arrival spacing.

ATC：Descend and maintain FL 240.

ATC：Descend at pilot's discretion, maintain FL 240.

ATC：Increase rate of descent.

ATC：Expedite descent to FL 270.

〔用語〕

rate of descent 降下率

〔訳例〕

ATCは、管制間隔を維持するために幾種類かの降下クリアランスを発行する。

ATC：Descend and maintain FL 240.

ATC：Descend at pilot's discretion, maintain FL 240.

ATC：Increase rate of descent.

ATC：Expedite descent to FL 270.

(7) 高度計のセット (Altimeter setting)

Pilots will calibrate altimeters by selecting QNH while leaving 14,000 feet. Local QNH is available through ATS or ATIS.

〔用語〕

calibrate 較正する

〔訳例〕

パイロットは、14,000ftを通過する時にQNHをセットして高度計を較正する。その地のQNHは、ATSまたはATISで入手できる。

(8) ホールディング (Holding)

ATC may instruct a pilot to hold for weather recovery or other purposes.

ATC：Spaceair 101, hold southeast of Hakodate VOR.

［用語］

hold 待機する　weather recovery 天候の回復

［訳例］

　　ATCは、パイロットに天候の回復またはその他の目的で待機の指示をすることがある。

　　ATC：Spaceair 101, hold southeast of Hakodate VOR.

第3節のまとめ英文 (The summary of section 3)

① The en-route course consists of the transition route, airway, and RNAV route. The transition route begins with the last fix of SID and ends at the intercepting point with the airway.

② Aircraft speed is expressed in IAS.

③ A pilot is instructed to change frequency along with control transfer. The pilot should report the present altitude with assigned altitude or flight level at initial contact.

④ Interpilot communication may be conducted on specified frequency. The company radio frequency should be used for inter-pilot communication of the aircraft stations of the same airline.

⑤ A pilot will request ATC course deviation for a case such as weather avoidance.

⑥ A pilot is requested to make PIREP when he/she has observed or encountered significant weather condition, in case of domestic operation.

第4節：到着の手順 (Arrival Procedures)

(1) アプローチとの通信設定 (Initial contact with approach)

　　A pilot should report the present altitude when establishing initial contact with approach control. A pilot is also requested to inform ATC of the ATIS code if it has been obtained.

> Pilot：Chitose Approach, Spaceair 101, leaving FL160 for 12,000, information C.

［用語］

initial contact 最初の通信設定　　approach control 進入管制

ATIS code ATIS のコード

［訳例］

　　パイロットは、進入管制と最初に通信設定する時には現在高度を報告すること。パイロットは、また、もしATISを入手していたならばその符号をATCに知らせるよう要求されている。

> Pilot：Chitose Approach, Spaceair 101, leaving FL160 for 12,000, information C.

(2) 速度制限 (Speed limit)

The airspeed is limited to 250kt at 10,000ft or below in the approach control area. It may be exceeded when permission to fly at a speed higher than 250kt has been granted by ATC.

> Pilot：Tokyo Control, Spaceair 101, request 320kt, we have a sick passenger on board.

［用語］

approach control area 進入管制区　exceed 超過する　grant 許可する

［訳例］

　　進入管制空域における対気速度は、10,000ft またはそれ以下では250ktに制限されている。ただし、ATCから250kt以上で飛行する許可を得ている場合には、高速で飛行できる。

> Pilot：Tokyo Control, Spaceair 101, request 320kt, we have a sick passenger on board.

(3) レーダー誘導 (Radar vectoring)

Arrival aircraft may be vectored to the traffic pattern or the final approach course.

① 誘導の開始 (Initiation of vectoring)

The controller gives the pilot the initial heading.

ATC：Spaceair 101, turn right heading 340 vector to Fukuoka VORTAC, maintain 7,000.

Pilot：Spaceair 101, right heading 340, maintain 7,000.

② 場周経路への誘導 (Vectoring to a traffic pattern)

ATC：Continue present heading vector to runway 22 traffic pattern.

Pilot：Spaceair 101, runway in sight, request visual approach.

③ 計器進入コースへの誘導 (Vectoring to an instrument approach course)

ATC：Spaceair 101, fly heading 060, vector to ILS Y runway 01R final approach course, descend and maintain 6,000.

Pilot：Heading 060, descend and maintain 6,000, Spaceair 101.

［用語］

arrival aircraft 到着機　　　　　traffic pattern 場周経路

final approach course 最終進入経路

visual approach 視認進入　　　　instrument approach 計器進入

［訳例］

到着機は、場周経路または最終進入経路へと誘導される。

① 誘導の開始

管制官は、パイロットに初期機首方位を与える。

ATC：Spaceair 101, turn right heading 340 vector to Fukuoka VORTAC, maintain 7,000.

Pilot：Spaceair 101, right heading 340, maintain 7,000.

② 場周経路への誘導

ATC：Continue present heading vector to runway 22 traffic pattern.

Pilot：Spaceair 101, runway in sight, request visual approach.

③　計器進入コースへの誘導

ATC：Spaceair 101, fly heading 060, vector to ILS Y runway 01R final approach course, descend and maintain 6,000.

Pilot：Heading 060, descend and maintain 6,000, Spaceair 101.

(4) ILS 進入 (ILS approach)

The ILS guides the pilot to precisely fly on the final approach course.

ATC：Spaceair 101, turn left heading 030, descend and maintain 3,000, 5 miles to YOTEI, cleared for ILS Y runway 01R approach.

Pilot：Left 030, descend and maintain 3,000, cleared for ILS Y runway 01R approach, Spaceair 101.

〔用語〕

precisely 精密に

〔訳例〕

ILSは、パイロットに最終進入コース上を精密に飛行するように誘導する。

ATC：Spaceair 101, turn left heading 030, descend and maintain 3,000, 5 miles to YOTEI, cleared for ILS Y runway 01R approach.

Pilot：Left 030, descend and maintain 3,000, cleared for ILS Y runway 01R approach, Spaceair 101.

(5) グラウンドコントロールアプローチ（GCA）

GCA controller vectors the pilot on the approach phase. GCA consists of two segments, the non-precision surveillance radar approach phase (ASR) and the precision approach phase (PAR). The controller vectors the pilot to the final approach course through voice communication.

ATC (ASR)：Turn right heading 180, prepare to descend, gear should be down.

ATC (PAR)：Spaceair 101, final controller, how do you read?

Pilot：Read you loud and clear.

ATC：Do not acknowledge further transmissions.

ATC：Turn right heading 180, maintain 1,500, 6 miles from touchdown, approaching glidepath.

ATC：Begin descent, 5 miles from touchdown.

ATC：Slightly right of course, turn left heading 178.

ATC：On glidepath.

ATC：Runway 18L, clear to land, wind 140 degrees 4 knots, on course, on glidepath.

〔用語〕

segment 区分　non-precision surveillance radar 非精密監視レーダー

〔訳例〕

　GCAのコントローラーは、進入段階にあるパイロットを誘導する。GCAは二つの区分から成り、一つは非精密監視レーダーの進入段階で、他は精密進入レーダーを利用する精密進入段階である。コントローラーは、音声通信を通じてパイロットを最終進入経路へと誘導する。

ATC (ASR)：Turn right heading 180, prepare to descend, gear should be down.

ATC (PAR)：Spaceair 101, final controller, how do you read?

Pilot：Read you loud and clear.

ATC：Do not acknowledge further transmissions.

ATC：Turn right heading 180, maintain 1,500, 6 miles from touchdown, approaching glidepath.

ATC：Begin descent, 5 miles from touchdown.

ATC：Slightly right of course, turn left heading 178.

ATC：On glidepath.

ATC：Runway 18L, clear to land, wind 140 degrees 4 knots, on course, on glidepath.

(6) 騒音軽減方式 (Noise abatement procedure)

The recommended procedures are ① preferential runway, ② preferential route, ③ reduced flap setting and ④ delayed flap operation.

〔用語〕

noise abatement procedure 騒音軽減方式

preferential runway 優先滑走路　　preferential route 優先経路

reduced flap setting　浅いフラップの使用

delayed flap operation フラップを遅らせる操作

〔訳例〕

　推奨されている騒音軽減方式は、①優先滑走路、②優先経路、③浅いフラップの使用および④フラップ使用を遅らせる方式である。

(7) 着陸のクリアランス (Landing clearance)

A landing clearance is issued by the tower during final approach. It provides weather information containing RVR when applicable, visibility, QNH, surface wind and other information such as reported windshear.

ATC：Spaceair 101, runway 01R cleared to land, visibility 2,500 meters, wind 350 degrees 5 knots.

Pilot：Runway 01R cleared to land, Spaceair 101.

〔用語〕

final approach 最終進入　　RVR 滑走路視距離　　visibility 視程

surface wind 地表風　　windshear ウィンドシア

110

[訳例]

　着陸のクリアランスは、最終進入中にタワーによって発行される。それは、RVR（該当する場合）、視程、QNH、地表風および報告されているウィンドシアのような他の情報を含む気象情報を提供する。

　　　ATC：Spaceair 101, runway 01R cleared to land, visibility 2,500
　　　　　　meters, wind 350 degrees 5 knots.
　　　Pilot：Runway 01R cleared to land, Spaceair 101.

第4節のまとめ英文 (The summary of section 4)

① STAR connects an arrival route to an approach course.

② A pilot is requested to inform ATC of the present altitude at initial contact with approach control. If the pilot has obtained ATIS information, its code should be informed as well.

③ The airspeed below 10,000ft in the approach control area is limited to 250kt. However, it may be exceeded if permission by ATC has been granted.

④ Arrival aircraft may be vectored to the traffic pattern or the final approach course.

⑤ ILS provides a pilot on the final approach course with precision approach guidance by means of radio signals, the localizer and the glide slope.

⑥ GCA controller guides the pilot to the normal approach paths through voice communication by calling out the aircraft deviation from the normal lateral and vertical approach paths.

⑦ Landing clearance is issued by the tower while an aircraft is on the final approach course. It includes the active landing runway, weather conditions and other necessary information.

第 5 節：緊急操作 (Emergency Procedures)

[1]　一般 (General)

(1) 緊急事態 (Emergency)

Emergency is a generic term meaning an aircraft encountered difficulties in flight.

[用語]

generic term 包括的な用語　encounter 遭遇する　difficulty 困難

[訳例]

　エマージェンシーは、航空機が運行中に困難に遭遇したことを意味する包括的な用語である。

(2) 緊急時の職権の実行 (Execution of emergency authority)

In an emergency requiring immediate action, a pilot may deviate from ATC clearance or instruction. He/she should promptly notify ATC of his/her emergency authority execution and obtain ATC instruction when normal operations are resumed.

[用語]

immediate action 即時の行動

emergency authority execution 緊急時職権の実行

[訳例]

　即座の行動を必要とする緊急状態においてパイロットは、ATCのクリアランスや指示から逸脱することがあり得る。パイロットは、ATCに緊急時の職権を実行したことを即座に通知し、通常状態に回復した時にはATCの指示を得ること。

(3) 緊急操作の表明 (Declaration of emergency)

The pilot should verbally declare an emergency or squawk transponder code 7700, and report the conditions with his/her intention.

〔用語〕

declaration 表明　　verbally 言葉により

squawk トランスポンダーを作動させる

〔訳例〕

　　パイロットは、緊急状態にあることを口頭で表明するかまたはトランスポンダーコード7700をセットして、緊急の状態とパイロットの意図するところを通報すること。

（4）使用周波数 (Frequencies to be used)

Distress and urgency call is made on the assigned frequency in use or on the emergency frequency 121.5MHz.

〔用語〕

assigned frequency in use 使用中の所定の周波数

〔訳例〕

　　遭難および緊急呼出しは、使用中の指定周波数または緊急周波数121.5MHzで実施される。

（5）航空機用救命無線機 (Emergency locator transmitter (ELT))

A compulsory aircraft station is equipped with Emergency Locator Transmitter (ELT). It transmits identification signal to the search and rescue satellite and tone signals to the rescue stations.

〔用語〕

emergency locator transmitter 航空機用救命無線機

compulsory aircraft station 義務航空機局

identification signal 識別信号

search and rescue satellite 捜索救難衛星　　rescue station 救難局

〔訳例〕

　　義務航空機局には航空機用救命無線機が装備されている。それは捜索救難衛星に識別信号を、またトーン信号を救難局に送信する。

［2］　遭難・緊急（Distress and urgency）

(1) 遭難呼出し（Distress call）

The distress call should be made in the following manner.

① MAYDAY preferably spoken three times for distress condition.

② The identification of the aircraft.

③ The distress call has absolute priority over all other communications. MAYDAY commands radio silence on the frequency in use.

［用語］

distress call 遭難呼出し

identification of the aircraft　航空機の識別符号

absolute priority 絶対的優先順位　　　　command 命令する

radio silence 通信の沈黙

［訳例］

遭難呼出しは、下記の方法で実施される。

① MAYDAY をなるべく3回前置する。

② 航空機の識別符号

③ 遭難呼出しは、他のすべての通信に対して絶対的優先順位を有する。MAYDAY は使用中の周波数に対し通信の沈黙を命じる。

(2) 緊急通信（Urgency communications）

The urgency communications should be made in the following manner.

① PAN-PAN preferably spoken three times for urgency condition.

② The identification of the aircraft.

③ The urgency communications have priority over all other communications, except distress, and all stations shall take care not to interfere with the transmission of urgency traffic.

114

[用語]

Interfere 妨害する

[訳例]

緊急通信は、下記の方法で実施される。

① PAN-PAN をなるべく 3 回前置する。

② 航空機の識別符号

③ 緊急通信は、遭難を除きすべての他の通信に優先権を有する。すべての局は、緊急通信の送信に混信を与えないように注意しなければならない。

(3) 遭難通報 (Distress messages)

The distress messages should be transmitted in the following manner.

① MAYDAY preferably spoken three times;

② name of the station addressed;

③ the identification of the aircraft;

④ the nature of the distress condition;

⑤ intention of the person in command;

⑥ present position, level, and heading.

[訳例]

遭難通報は、下記の方法で送信される。

① MAYDAY をなるべく 3 回前置；

② 相手局の名称；

③ 航空機の識別；

④ 遭難の種類；

⑤ 機長の意図；

⑥ 位置、高度、機首方位

(4) 緊急通報 (Urgency messages)

The urgency messages should be transmitted in the following manner.

① PAN-PAN preferably spoken three times;

② name of the station addressed;

③ the identification of the aircraft;

④ the nature of the urgency condition;

⑤ intention of the person in command;

⑥ present position, level, and heading;

⑦ any other useful information.

［訳例］

　　緊急通報は、下記の方法で送信される。

① PAN-PAN をなるべく 3 回前置；

② 相手局の名称；

③ 航空機の識別；

④ 緊急事態の種類；

⑤ 機長の意図；

⑥ 位置、高度、機首方位；

⑦ その他の有用な情報 .

（5）　地上局の措置

① Subject aircraft call sign;

② Call sign of the ground station;

③ Roger;

④ MAYDAY

［訳例］

　　緊急通報は、下記の方法で送信される。

① 遭難通報を送信した航空機局の呼出名称；

② 自局の呼出名称；

③ 了解；

④ 遭難；

[3] 他の緊急事項（Other items concerning an emergency）
(1) 燃料放出 (Fuel dumping)

A pilot may have to dump fuel to reduce gross weight prior to landing. He/she should inform ATC of required duration, and obtain instructions for the area.

> Pilot : Departure, Spaceair 101, request fuel dump for about 10 minutes, then return to Tokyo.
>
> ATC : Spaceair 101, Departure, roger, fly heading 120, vector for dumping area, maintain 6,000, start dumping, report completion.

〔用語〕

fuel dumping 燃料放出　　gross weight 総重量

〔訳例〕

　パイロットは、着陸に先立って総重量を軽減するために燃料を放出することがある。パイロットは、ATCに必要時間を知らせかつ実施場所について指示を得なければならない。

> Pilot : Departure, Spaceair 101, request fuel dump for about 10 minutes, then return to Tokyo.
>
> ATC : Spaceair 101, Departure, roger, fly heading 120, vector for dumping area, maintain 6,000, start dumping, report completion.

(2) 通信機能故障 (Communication failure)

When communication establishment is impossible, a pilot should attempt to contact with other station. When this attempt fails, he/she should squawk 7600, and inform ATC of position, altitude, and his/her intention with "blind transmission" .

> Pilot : Tokyo Control, Spaceair 101, transmitting in blind, over Yankee, 1234, FL150, Oscar 1249, Temple next. Spaceair

101.

［用語］

squawk トランスポンダーコードをセットすること

blind transmission 一方送信

［訳例］

　通信設定が出来ない時には、パイロットは他の局との通信設定を試みる。もしそれも駄目ならば、パイロットはトランスポンダー 7600 をセットし、ATC にその位置、高度およびパイロットの意図するところを一方送信する。

　　Pilot：Tokyo Control, Spaceair 101, transmitting in blind, over
　　　　　 Yankee, 1234, FL150, Oscar 1249, Temple next. Spaceair
　　　　　 101.

(3) 燃料の欠乏（Minimum fuel）

　The pilot in command shall advise ATC of a minimum fuel state by declaring MINIMUM FUEL when, having committed to land at a specific aerodrome, the pilot calculates that any change to the existing clearance to that aerodrome may result in landing with less than planned final reserve fuel.

［用語］

planned final reserve fuel 飛行計画の根拠となる燃料の量

［訳例］

　機長は、特定の空港に向かっている場合、現在出されている管制許可に変更があれば、着陸の時点の燃料が、飛行計画の根拠となる燃料の量未満となると予想されるときは、管制機関に対してMINIMUM FUELと宣言しなければならない。

第５節のまとめ英文（The summary of section 5）

① In an emergency requiring immediate action, a pilot may deviate from ATC clearance or instruction. He/she should promptly notify ATC of his/her emergency authority execution and obtain ATC instruction when normal operations are resumed.

② The pilot should verbally declare an emergency or squawk transponder code 7700, and report the conditions with his/her intention.

③ Distress and urgency call is made on the assigned frequency in use or on the emergency frequency, 121.5MHz.

④ A compulsory aircraft station is equipped with Emergency Locator Transmitter（ELT）. It transmits identification signals to the search and rescue satellite and tone signals to the rescue stations.

⑤ The distress call has absolute priority over all other communications. MAYDAY commands radio silence on the frequency in use.

⑥ When communication establishment is impossible, a pilot should attempt to contact with other station. When this attempt fails, he/she should squawk 7600, and inform ATC of position, altitude, and his/her intention with "blind transmission".

⑦ The pilot in command shall advise ATC of a minimum fuel state by declaring MINIMUM FUEL when, having committed to land at a specific aerodrome, the pilot calculates that any change to the existing clearance to that aerodrome may result in landing with less than planned final reserve fuel.

第6章

通信・航法装置
Communication Navigation Systems・Equipment

この章では無線通信システムおよび通信機器の概要について述べ、関連英語の習得に資することとする。

This chapter intends to promote understanding of English related to radio system operations by presenting brief information on communication systems and equipment.

第1節：通信システム（Communication System）

Communication system consists of power supply, transmitting system and receiving system.

The power for communication system is usually a low voltage direct current（DC）.

The transmitting system produces high frequency current, the carrier wave, by an oscillator and a modulated carrier wave by modulating the carrier wave with the signal wave. The modulated carrier wave is changed into radio wave by the transmitting antenna. The transmitting antenna emits radio waves into space. Radio waves are propagated in space and received by a receiving antenna. The receiving antenna changes the radio wave into electric current. It is demodulated to select signal wave. The signal wave is fed to a display unit as a voice signal, video signal, or digital data.

〔用語〕

high frequency current 高周波電流	carrier wave 搬送波
oscillator 発振器	modulated carrier wave 変調波
signal wave 信号波	radio wave 電波
demodulate 復調する	display unit 表示器
video signal 映像信号	voice signal 音声信号

〔訳例〕

　通信システムは、電源、送信装置及び受信装置から成る。電源は、通常低電圧の直流である。送信装置は、発振器により高周波電流（搬送波）を発生し、また搬送波を信号波で変調することにより変調搬送波を発生する。　変調搬送波は、送信アンテナで電波に変換される。送信アンテナは、それを空間に放射する。電波は、空間を伝搬し、受信アンテナにより受信され、電流に変換され、復調されて通信信号が音声信号、映像信号またはデジタルデータとして表示装置に送られる。

（1）電源（Power supply）

　Airliners' main power supply is usually supported by engine driven AC generators. Airborne radio equipment is operated on DC. AC power is converted to DC by a rectifier and the voltage is changed to the appropriate volt for radio equipment. The transformer rectifier unit（TRU）in the power distribution system adjusts the voltage and rectifies AC into DC. TRU contains a smoothing circuit to eliminate AC component remaining in rectifier output.

〔用語〕

main power supply 主電源

engine driven generator エンジン駆動の発電機

rectifier 整流器　　transformer 変圧器

power distribution system 分電系統　　smoothing circuit 平滑回路

output 出力

［訳例］

　定期旅客機の主電源は、通常エンジン駆動の交流発電機によって賄われている。機上の無線装備は、直流で作動する。交流電力は整流器によって直流に転換され、電圧は無線設備に適切な電圧に変換される。分電系統の変圧・平滑装置（TRU）は、電圧を調整し、交流を直流に変換する。TRUに内蔵される平滑回路は、整流器出力に残存する交流成分を除去する。

(2) 送信装置 (Transmitting system)

① 　The oscillator in a transmitter generates high frequency carrier wave. The modulator modulates the carrier wave by signal wave through amplitude modulation (AM), frequency modulation (FM) or pulse modulation (PM) method. The modulated carrier wave is fed to the transmitting antenna through a feeder.

② 　The feeder conducts high frequency current between the transmitter and the antenna. It contributes impedance matching between the transmitter and the antenna. The coaxial cable is a feeder used for frequencies up to UHF, and the waveguide is used for microwaves.

③ 　The transmitting antenna converts high frequency current into a radio wave, and emits it into space.

［用語］

oscillator 発振器	carrier wave 搬送波
modulator 変調器	signal wave 信号波
amplitude 振幅	AM 振幅変調　　　FM 周波数変調
PM パルス変調	modulated carrier wave 変調搬送波
feeder 給電線	impedance インピーダンス
matching マッチング（整合）	coaxial cable 同軸ケーブル
waveguide ウエーブガイド（導波管）	

microwave マイクロ波（極超短波）　　　space 空間

［訳例］

① 送信機内の発振器は、高周波搬送波を発生させる。変調器は、搬送波を振幅変調、周波数変調またはパルス変調方式によって、信号波で変調する。変調搬送波は給電線を介して送信アンテナに送られる。

② 給電線は、送信機とアンテナの間における高周波電流を伝導する。それは送信機とアンテナとの間のインピーダンスマッチングを保つ。同軸ケーブルは、UHF までの周波数に使用される給電線であり、ウエーブガイドは、マイクロ波に使用される給電線である。

③ 送信アンテナは、高周波電流を電波に変換し、空間に放射する。

（3）電波の伝搬（Propagation of radio waves）

Radio waves show various characteristics such as penetration, diffraction, reflection, and attenuation in relation to the area and frequencies.

① 地上波（Ground waves）

Near the surface, radio waves are attenuated. Radio waves of their frequencies higher than VHF attenuate more. The propagation of VHF and higher frequency radio waves is limited to the "line of sight".

Radio waves of their frequencies lower than MF propagate far distance. MF has been used for radio broadcasting.

Radio waves traveling along the surface are diffracted and reach the distance farther than the "line of sight".

② 対流圏波（Tropospheric waves）

In the troposphere, weather conditions adversely affect high frequency radio waves, VHF and higher, resulting in fading.

③ 電離層波（Ionospheric waves）

The ionosphere reflects HF radio waves. VHF and higher

frequency radio waves penetrate the ionosphere. HF radio waves keep reflecting between the ionosphere and the surface for long distance propagation. Intense ionization density causes Dellinger phenomenon or magnetic storm.

［用語］

propagation 伝搬（伝わるということ）　　characteristic 特性、特徴

diffraction 回折　　　　　reflection 反射

attenuation 減衰　　　　　line of sight 見通し距離

troposphere 対流圏　　　　fading フェイディング

ionosphere 電離層　　　　ionization 電離

Dellinger phenomenon デリンジャー現象

magnetic storm 磁気嵐

［訳例］

　電波は、場所と周波数とのかかわりで貫通、回折、反射および減衰などのさまざまな特徴を示す。

①　地表面近くでは、電波は減衰する。VHFより高い周波数の電波はより多く減衰する。VHFおよびより高い周波数の電波の伝搬は、見通し距離内に制限される。

　　MFより低い周波数の電波は、遠くまで伝搬する。MFは、ラジオ放送に使用されている。地表に沿って伝搬する電波は、回折され見通し距離より遠くまで到達する。

②　対流圏においては、気象状態が周波数の高い電波に不利に影響し、VHFおよびより高い周波数の電波はフェイディングを起こす。

③　電離層は、HFの電波を反射する。VHFおよびより高い周波数の電波は、電離層を貫通する。HFの電波は、電離層と地表面の間の反射を繰り返し、遠くまで伝搬する、強度のイオン化は、デリンジャー現象や磁気嵐を発生させる。

(4) 受信装置 (Receiving system)

An antenna receives a radio wave that resonates with the antenna's characteristic frequency. The antenna converts received radio wave into an electric current, and apply it to the receiver. The receiver demodulates the received current to select signal wave, which is formed into audio signal, visual signal or digital data and displayed on the display unit, or it is applied to the data processing system.

［用語］

resonate 共鳴する　　　　　　　characteristic frequency 固有周波数

demodulate 復調する　　　　　　signal wave 信号波

audio signal 音声信号　　　　　　visual signal 映像信号

data processing system データ処理装置

［訳例］

　アンテナは、その固有周波数に共鳴する電波を受信する。アンテナは、受信した電波を電流に転換し、受信機に与える。受信機は、信号波を取り出すために受信電流を復調する。そしてそれは音声信号、映像信号またはデジタルデータにされて表示装置に表示され、またはデータプロセスシステムに提供される。

第 1 節のまとめ英文 (The summary of section 1)

①　The oscillator generates carrier wave. The modulator modulates carrier wave by signal wave through AM, FM, or PM method.

②　The feeder contributes impedance matching. The coaxial cable is applied to the frequencies up to UHF, and the waveguide for microwaves.

③　The transmitting antenna converts high frequency current into a radio wave and emits it into space.

④　Radio waves show various characteristics such as penetration, diffraction, reflection, and attenuation during propagation.

⑤　VHF and higher frequency radio waves propagation is limited to the "line of sight" near the surface. MF propagates far distance. MF has been used for radio broadcasting.

⑥　In the troposphere, weather conditions adversely affect high frequency radio waves, VHF and higher, resulting in fading.

⑦　The ionosphere reflects HF radio waves. VHF and higher frequency radio waves penetrate the ionosphere. HF radio waves keep reflecting between the ionosphere and the surface for long distance propagation. Intense ionization density causes Dellinger phenomenon or magnetic storm.

第 2 節：レーダーの特徴 (Typical Features of a Radar)

(1) 周波数 (Frequencies)

Frequencies used for radar are high because detection ability is higher with high frequency radio waves. However, they attenuate more. Detection ability and coverage compromises each other.

〔用語〕

detection ability 探知能力　　　coverage 覆域

〔訳例〕

　　レーダーの使用周波数は、高い周波数である。高い周波数のほうが探知能力が優れているからである。ただし高い周波数の電波は、減衰が大きい。探知能力と覆域は互いに妥協し合う。

(2) 変調方式 (Modulation method)

Pulse modulation method is employed to keep an interval, the reception period, between transmissions of each pulse.

126

pulse modulation method パルス変調方式 　　　interval 間隔

reception period 受信期間

［訳例］

　各々のパルスの送信と送信の間に受信時間帯を設けるために、パルス変調方式が利用されている。

(3) レーダーの種類 (Types of radar system)

There are two types of radar system, the primary radar and the secondary radar.

① 　The primary radar transmits a radio wave into wide range, and an object reflects the radio wave, which is received by the original station's antenna. The azimuth of the target is synchronized with the direction of the antenna, and the distance to the target is in proportion to the elapsed time of the radio wave spent for both ways between the station and the target. Radars used for ATC such as ASR and ARSR are primary radars.

② 　The secondary radar transmits a specific radio wave to a specific station, and the specific station retransmits a specific radio wave, which is received by the original station to acquire required information. The original transmitting station is regarded as an interrogator, and the retransmitting station as a transponder. DME is an example of the secondary radar system.

［用語］

primary radar 一次レーダー	secondary radar 二次レーダー
azimuth 方位	target 目標物
synchronize 同期する	in proportion to 〜と比例して
elapsed time 経過時間	retransmit 再送信する
interrogator 質問機	transponder 応答機

［訳例］

一次レーダーと二次レーダーの二つの型式のレーダーシステムがある。

① 　一次レーダーは、電波を広範囲に送信し、対象物が電波を反射し、反射された電波は発信局のアンテナによって受信される。目標物の方位は、アンテナの方位と同調している。また目標物までの距離は、電波が局と目標物との間の往復に経過した時間に比例している。ATC に使用されている ASR や ARSR は、一次レーダーである。

② 　二次レーダーは、特定の電波を特定の局に対して送信し、その特定の局はまた特定の電波を再送信する。再送信された電波は発信局によって受信され、必要な情報が入手される。発信局は質問機とみなされ、再送信する局は応答機とみなされる。DME は、二次レーダーの例である。

第 2 節のまとめ英文 (The summary of section 2)

① 　Frequencies used for radar are high because detection ability is higher with high frequency radio waves. However, detection ability and coverage compromises each other.

② 　Pulse modulation method is employed to keep receiving period between transmissions of each pulse.

③ 　The primary radar transmits a radio wave into wide range, and an object reflects the radio wave, which is received by the original station's antenna. The azimuth of the target is synchronized with the direction of the antenna, and the distance to the target is in proportion to the elapsed time of the radio wave spent for both ways between the station and the target.

④ 　The secondary radar transmits a specific radio wave to a specific station, and that station retransmits a specific radio wave, which is received by the original station. The original transmitting station is regarded as an interrogator, and the retransmitting station as a transponder. DME is an example of the secondary radar system.

第3節：レーダー装備（Radar Equipment）

（1）空港内地表面探知装置（ASDE）マルチラテレーションシステム（MLAT）

ASDE is a primary radar system to monitor traffic in the maneuvering area of an airport. It is operated on SHF for high detection ability in a short range.

The multilateration is introduced to eliminate blind areas of the ASDE.

This system is used for the airport surface surveillance of mode S transponder equipped aircraft.

〔用語〕

primary radar system 一次レーダー　　maneuvering area 走行区域
detection ability 検出能力

〔訳例〕

ASDEは、飛行場内の走行区域における交通を監視するための一次レーダーである。それは、小範囲内における高検出能を必要とするために、SHFで作動する。

ASDEのブラインドエリアを解消するために、マルチラテレーションシステムが導入されている。

このシステムは、モードSトランスポンダーを搭載した航空機の飛行場面における監視に利用されている。

(2) 空港監視レーダー (ASR)

ASR is a primary radar system to monitor aircraft departing from or arriving at an airport. Aircraft altitude information is added by SSR. ASR is operated on UHF.

［用語］

altitude information 高度情報

［訳例］

ASRは、飛行場における出発機および到着機を監視するための一次レーダーである。航空機の高度情報は、SSRから追加される。ASRは、UHFで作動する。

(3) 航空路監視レーダー (ARSR)

ARSR is a primary radar system to monitor aircraft on the domestic airways. It is operated on UHF for greater range capability.

［用語］

domestic airway 国内航空路　　　range capability 覆域性能

［訳例］

ARSRは、国内航空路上の航空機を監視するための一次レーダーである。それは長距離覆域性能を必要とするので UHF で作動する。

(4) 洋上航空路監視レーダー (ORSR)

ORSR is a secondary radar system to monitor aircraft on the oceanic routes. It is operated on UHF.

［用語］

secondary radar system 二次レーダー　　　oceanic route 洋上航空路

［訳例］

ORSRは、洋上航空路上の航空機を監視するための二次レーダーである。それはUHFで作動する。

(5) 精密進入レーダー (PAR)

PAR is a primary radar system that is used by a GCA controller to vector aircraft on the final approach course for precision approach. It is operated on SHF.

［用語］

vector aircraft 航空機を誘導する　final approach course 最終進入経路 precision approach 精密進入

［訳例］

　PARは、GCAコントローラーが、最終進入経路にいる航空機を精密進入のために誘導するに当たって使用する一次レーダーである。それはSHFで作動する。

(6) 二次監視レーダー (SSR)

　SSR is a secondary radar system that is operated on UHF. It is a ground installed interrogator and works in conjunction with an airborne transponder. The SSR is co-located with ASR or ARSR to display information on ASR or ARSR radar screen. SSR mode-S is capable of individual aircraft addressing and data link communication.

［用語］

interrogator 質問機　　　in conjunction with ～とともに
airborne 機上の　　　　　transponder 応答機
individual aircraft addressing 各機個別呼出し

［訳例］

　SSRは、UHFで作動する二次レーダーである。それは地上装備の質問機であって機上装備の応答機とともに機能する。SSRは、その情報をASRやARSRのスクリーン上に表示するためにASRやARSRと併置されている。SSRのモードSは、各航空機別の個別呼出しとデータリンク通信機能を持っている。

(7) 機上衝突防止装置 (ACAS)

　Airborne collision avoidance system is called ACAS by ICAO and TCAS by the manufacturer. Mode-S transponder operates ACAS/

TCAS on UHF. Interrogation signal is transmitted from the mode-S transponder equipped aircraft into the hazardous area of collision, which is received by an aircraft within the specified area, and then a reply signal is retransmitted to indicate existence of danger by activating traffic advisory (TA) or resolution advisory (RA). Closure rate of nearing aircraft is calculated by altitude and distance information of the hazardous aircraft.

〔用語〕

collision avoidance system 衝突防止装置　　manufacturer 製造者
interrogation signal 質問信号

hazardous area 危険域　　　　　　specified area 所定空域

reply signal 応答信号　　　　　　traffic advisory 位置情報

resolution advisory 回避情報　　closure rate 接近度

〔訳例〕

　機上衝突防止装置は、ACAS（ICAO名称）またはTCAS（製造者名称）と呼ばれている。Mode-S ATCトランスポンダーは、UHFでACAS/TCASを作動させる。Mode-S ATCトランスポンダー装備の航空機から衝突危険空域に対して質問信号が発信され、所定空域内にいる航空機によって受信され、応答信号が送信される。応答信号は、位置情報または回避情報を作動させることによって危険を知らせる。近づきつつある航空機の接近度は、その危険な航空機の高度情報と距離情報によって算出される。

(8) 距離測定装置（DME）

　DME is a secondary radar system to measure slant distance from an aircraft to a reference point of a ground station. The system consists of an airborne interrogator and a ground installed transponder that works in conjunction with VOR or ILS by means of UHF. DME coverage is equivalent to or more of VOR or ILS coverage.

The transponder transmits reply signal on the frequency different from the interrogation signal and with specified time delay after receiving interrogation pulses. The elapsed time between transmission and reception of the signals is converted into distance and displayed in the cockpit.

[用語]

slant distance 斜距離	reference point 標点
in conjunction with 〜とともに	interrogation signal 質問信号
specified time delay 所定の時間間隔	elapsed time 経過時間

[訳例]

　DMEは、航空機から地上局の標点までの斜距離を測定するための二次レーダーである。DMEは、機上の質問機と地上装備の応答機から成り、VORやILSと連動してUHFで作動する。DMEの到達範囲はVORやILSと同等またはそれ以上である。

　応答機は、質問機の周波数とは異なる周波数で、かつ、質問パルスの受信後所定の時間を置いて応答信号を送信する。送信・受信の間の経過時間を距離に変換し、コックピットに表示する。

(9) 低高度用電波高度計 (LRRA)

LRRA is a primary radar system that is operated on a frequency modulated SHF radio wave. It measures distance to the object directly below the aircraft between 2,500ft and zero feet. It provides pilots with altitude information for precision approach and landing maneuver. The elapsed time of the radio wave emitted from the aircraft and reflected by an object on the surface and is received by the aircraft is equivalent to the vertical distance.

[用語]

frequency modulated SHF 周波数変調 SHF

precision approach and landing maneuver 精密進入着陸時の操縦操作

vertical distance 縦の距離

［訳例］

　　LRRAは、周波数変調のSHFにより作動する一次レーダーである。それは航空機真下の物体までの距離を測定するが、その範囲は零フィートから2,500フィートまでである。それは精密進入着陸時の操縦操作のために必要な高度情報をパイロットに提供する。航空機から発射された電波が地上の物体によって反射され、そして航空機によって受信されるまでの経過時間は、垂直距離に等しい。

（10）機上気象レーダー（Airborne weather radar）

　　Airborne weather radar is a primary radar system that is operated on SHF with the maximum coverage of approximately 300nm. It is used to detect significant weather, traffic conflict and terrain information. Stabilization of transmitting radio wave against pitching and rolling of the aircraft is controlled by IRS signals.

［用語］

maximum coverage 最大到達範囲　　significant weather 悪天候

traffic conflict 危険な交通　　terrain information 地形情報

stabilization 安定

［訳例］

　　機上気象レーダーは、最大到達範囲約300マイルのSHFで作動する一次レーダーである。それは悪天候、危険な交通および地形情報を検知するために使われる。送信電波の機体の縦揺れや横揺れに対する安定は、IRSからの信号によって制御される。

第3節のまとめ英文（The summary of section 3）

①　ASDE is a primary radar system to monitor traffic in the maneuvering area. It is operated on SHF for high detection ability in a short range.

The multilateration is introduced to eliminate blind areas of the ASDE.

This system is used for the airport surface surveillance of mode S transponder equipped aircraft.

② ASR is a primary radar system to monitor aircraft departing from or arriving at an airport. Aircraft altitude information is added by SSR. ASR is operated on UHF.

③ ARSR is a primary radar system to monitor aircraft on the domestic airways. It is operated on UHF for greater range capability.

④ ORSR is a secondary radar system to monitor aircraft on the oceanic routes. It is operated on UHF.

⑤ PAR is a primary radar system that is used by a GCA controller to vector aircraft on the final approach course for precision approach. It is operated on SHF.

⑥ SSR is a secondary radar system that is operated on UHF. It is a ground installed interrogator and works with an airborne ATC transponder. The SSR is co-located with ASR or ARSR to display information of the primary radar and the secondary radar on ASR or ARSR radar screen. SSR mode-S is capable of individual aircraft addressing and data link communication.

⑦ Mode-S transponder operates ACAS/TCAS on UHF. Interrogation signal is transmitted from the mode-S transponder equipped on an aircraft into the hazardous area of collision, which is received by an aircraft within the specified area, and then a reply signal is retransmitted to indicate existence of danger by activating traffic advisory (TA) or resolution advisory (RA) . Closure rate is calculated by altitude and distance information of the hazardous aircraft.

⑧　DME is a secondary radar system to measure slant distance from an aircraft to a ground station. The system consists of an airborne interrogator and a ground installed transponder that works with VOR or ILS on UHF. DME coverage is equivalent to or more of VOR or ILS coverage.

The elapsed time between transmission and reception of the signals is converted into distance and displayed in the cockpit.

⑨　LRRA is a primary radar system that is operated on a frequency modulated SHF radio wave. It measures distance to the object directly below the aircraft between 2,500ft and zero feet. It provides altitude information for precision approach and landing.

⑩　Airborne weather radar is a primary radar system that is operated on SHF with the maximum coverage of approximately 300nm. It is used to detect significant weather, traffic conflict and terrain information. Stabilization of transmitting radio wave is controlled by IRS signals.

第４節：無線航法システム（Radio Navigation System）

（1）マーカービーコン（Marker beacon）

A marker beacon is a radio beacon to inform an aircraft of a specific point that it is flying over. Marker beacons are installed on an airway and along the ILS final approach course. A marker beacon vertically emits VHF fan type radio wave that is modulated by an audible frequency.

An aircraft flying over the point where a marker beacon is installed on an airway will receive the beacon signals, and the pilot will recognize that he/she is flying over the specific point by monitoring indicator light and aural signals. Along the ILS final approach course, the outer marker that indicates the final approach point, the middle marker that indicates CAT Ⅰ decision altitude point, and the inner marker that indicates CAT Ⅱ decision height point are installed.

However, recent expansion of CNS/ATM contributes conversion of marker beacons that are installed on airways into VOR/DME or GPS and the markers located along the ILS final approach course into positioning information of DME or DGPS.

〔用語〕

radio beacon 無線標識　　　specific point 特定の地点

airway 航空路　　　　　　final approach course 最終進入コース

fan type radio wave 扇形電波　audible frequency 可聴周波数

recognize 認める　　　　　monitor モニターする

indicator light 指示灯　　　aural signal 音の信号

final approach point　最終進入開始点

decision height point 決心高の地点

decision altitude point 決心高度の地点

〔訳例〕

　マーカービーコンは、上空を飛行する航空機に対し特定の地点を知らせる無線標識である。それは航路上およびILSの最終進入コース上に設置される。マーカービーコンは、可聴周波数で変調されたVHFの扇形電波を上空に発射する。

　航路上のマーカービーコン設置点上空を飛行する航空機は、マーカービーコンの信号を受信し、パイロットは、表示灯と音の信号によって特定地点の上空飛行を確認する。ILSの最終進入コース上では、進入開始点を示すアウターマーカー、CAT I ILSアプローチの決心高度

の地点を示すミドルマーカーおよびCATⅡILSアプローチの決心高の地点を示すインナーマーカーが設置される。

　しかしながら最近のCNS/ATMの発展に伴い、航路上のマーカービーコンは、VOR/DMEやGPSに、またILSコース上のマーカーは、DMEやDGPSによる測位情報に転換しつつある。

(2) NDB/ADF

NDB/ADF is a system to find location and direction of a radio beacon station. NDB is a ground station that emits LF or MF radio signals and ADF is an airborne receiver. Some factors such as night effect or interference of similar frequency radio waves nearby will adversely affect the range or the accuracy of directional information of the system.

〔用語〕

radio beacon station 無線標識局　　night effect 夜間効果

similar frequency radio wave 類似周波数の電波

accuracy 精度　　　　　　　　　directional information 方位情報

〔訳例〕

　NDB/ADFは、無線標識局の位置およびその方位を求めるためのシステムである。NDBは、LFまたはMFを放射する地上局で、ADFは機上の受信機である。夜間効果や近くの類似した周波数の電波との干渉などの要素は、このシステムの到達範囲や方位情報の精度に悪影響をもたらす。

(3) VOR

A VOR station provides pilots with bearing information by emitting VHF radio signals, a reference phase signal and a variable phase signal. The aircraft VOR receiver detects phase difference of the two signals and determines the bearing to the station. DME is co-located at VOR stations.

Domestic airways are established depending on the radio signals of VOR stations. Airways connecting VOR stations are VOR routes, and the ones referring to VOR stations are RNAV routes based on VOR/DME.

[用語]

bearing information 方位情報　　reference phase signal 基準位相信号
variable phase signal 可変位相信号　　phase difference 位相差

[訳例]

　VORは、基準位相信号と可変位相信号とから成るVHFの電波を放射してパイロットに方位情報を提供する。機上のVOR受信機は、二つの信号の位相差を探知して、航空機のVOR局に対する方位を決定する。VOR局にはDMEが併置されている。

　国内航空路は、VORの電波を基に設定されている。VOR局を結んだ航空路は、VORルートであり、VOR局を参照にしているルートは、VOR/DMEを基準とするRNAVルートである。

(4) ILS

ILS is a system to precisely guide an aircraft on the final approach to landing by means of radio waves that provide lateral and vertical guidance. The lateral guidance radio wave is the Localizer (LLZ) operated on VHF, and the vertical guidance radio wave is the Glide path (GP) operated on UHF which is channeled to the localizer frequency. Deviation of an aircraft position from LLZ or GP is indicated in the cockpit and the pilot corrects the path to LLZ and GP to fly on the designated approach to landing path.

The location of the final approach point, CAT-I decision altitude point and CAT-II decision height point are specified, and a pilot acknowledges them by means of either marker beacons, ILS DME, barometric altimeter or, radio altimeter.

［用語］

　final approach 最終進入

　lateral and vertical guidance 平面および縦の誘導

　lateral guidance radio wave 平面誘導電波

　designated approach to landing path 所定の進入着陸経路

　final approach point 最終進入開始点

　decision height point 決心高地点

［訳例］

　　ILSは、航空機を平面および縦面の誘導電波によって精密に最終進入着陸へと誘導するためのシステムである。平面の誘導電波はVHFで作動するLLZで、縦面の誘導電波はUHFで作動しかつLLZ周波数にチャネリングしているGPである。LLZまたはGPからの航空機の変位は、コックピットに指示されていて、パイロットは所定の進入着陸経路を飛行するように自機の経路を修正する。

　　最終進入開始点、CAT-1の決心高度点およびCAT-Ⅱの決心高点は指定されていて、パイロットはそれらの地点をマーカービーコン、ILS DME、気圧高度計または電波高度計によって確認する。

第4節のまとめ英文 (The summary of section 4)

①　Marker beacons are installed on an airway and an ILS final approach course to provide over-flying aircraft with the specific point where the beacon is located. The outer marker (OM) is located at the final approach point of ILS, the middle marker (MM) is located at the point of CAT Ⅰ decision altitude and the inner marker (IM) at the point of CAT Ⅱ decision height. Pilots identify those specific points by means of indicating light and audible signals when they fly over a marker beacon.

②　NDB/ADF is a system to find location and direction of a radio beacon station. NDB is a ground station and ADF is an airborne receiver. Some factors such as night effect or interference of similar frequency radio waves nearby will adversely affect the range or the accuracy of directional information.

③ A VOR station provides pilots with bearing information by emitting VHF radio signals, a reference phase signal and a variable phase signal. The aircraft VOR receiver detects phase difference of the two signals and determines the bearing to the station. DME is co-located at VOR stations.

④ ILS precisely guides an aircraft on the final approach by means of radio waves that provide lateral and vertical guidance. The lateral guidance radio wave is the Localizer (LLZ) operated on VHF, and the vertical guidance radio wave is the Glide path (GP) operated on UHF which is channeled to the localizer frequency. Deviation of an aircraft position from LLZ or GP is indicated in the cockpit and the pilot corrects the path to LLZ and GP to fly on the designated approach to landing path.

The location of the final approach point, CAT-I decision altitude point and CAT-II decision height point are specified, and a pilot acknowledges them by means of either marker beacons, ILS DME, barometric altimeter or radio altimeter.

第 5 節：衛星通信・航法システム
(Satellite Communication Navigation System)

(1) 国際移動通信衛星機構 (IMSO)

The organization originally established for providing the maritime communication service, and managed under the international agreement. The extent of the service eventually has been developed to cover aircraft and ground mobile operations using the satellites name of INMARSAT. By the amendment of the agreement , the operational section has been transferred to a private sector (Inmarsat Corp.)

[用語]

international agreement 　　国際協定

amendment of the agreement 　　条約改正

[訳例]

　当初は国際船舶通信を対象にできた組織で、国際協定のもとに運営されていた。その後航空通信及び陸上移動通信についても衛星を提供するようになり、その衛星もINMARSATと呼ばれている。条約改正により民間（Inmarsat 社）に移管されて引き続き事業を運営している。

(2) 全地球測位システム（GPS）

　GPS has a greater accuracy than the conventional navigation systems and most receivers are capable of simultaneously receiving and using during high-speed transportation anywhere on the earth. There are six orbital planes containing at least four satellites on each and equally spaced 20,000 km above the surface. GPS system consists of 24 satellites or more that orbit the earth and the supporting unit on the ground. When required number of satellites are received, highly accurate position can be obtained anywhere on the earth. The position information provided by GPS has a higher accuracy than of VOR/DME. However, when the signals are used throughout all flight phases, differential GPS is required due to lack of the availability and continuity.

[用語]

conventional navigation systems 従来の航法システム

continuity 継続性

142

［訳例］

　GPSは、従来の航法システムに比べて精度が高いこと、地球上のどこでも高速移動中であっても多数の利用者が同時に利用可能なことなどの特徴がある。GPSは高度約20,000キロメートルの6つの周回軌道に各々最低4個、合計24個以上配置された人工衛星と衛星を維持管理する地上システムにより構成されている。GPSは位置算出に必要な数の衛星を受信できるところであれば、地球上どこでも高い精度の位置情報を得ることができる。GPSが提供する位置の精度はVOR/DMEよりも高いが、利用可能性やサービスの継続性の問題があり、すべての飛行フェーズにわたってGPSを使用するには、GPS補強信号が必要である。

（3）運輸多目的衛星（MTSAT）

　Due to interoperability with Inmarsat in terms of aeronautical satellite communication safety service, MTSAT made a contribution as an international safety infrastructure in Asia Pacific region to overseas air traffic control agencies and airlines. It worked for navigation and surveillance as well as communication to provide SBAS (Satellite-Based Augmentation System), which supports safer and securer navigation for oceanic flights, in particular, to improve accuracy and reliability of GPS information, by delivering measurement errors and corrections, and augmentation through satellites. In 2020, its aeronautical satellite solutions were terminated. SBAS function is being succeeded by QZSS (Quasi-Zenith Satellite System).

［用語］

Interoperability 相互運用性　　　contribution 貢献

infrastructure インフラ（社会的基本施設）　　　surveillance 監視機能

QZSS (Quasi-Zenith Satellite System) 準天頂衛星みちびき

［訳例］

　　MTSATの航空衛星通信サービスはInmarsatと相互運用性を持ち、海外の管制機関や航空会社も利用できるアジア太平洋地域の航空交通インフラとして貢献した。さらに通信サービスだけでなく航法・監視機能も有し、GPS衛星で得られる測位情報をより高い精度や高信頼にするための誤差・補強情報を衛星経由で航空機により安全で確実な航法となるSBAS(Satellite-Based Augmentation System)を提供したが、航空衛星サービスは2020年で終了し、SBAS機能は準天頂衛星みちびき（QZSS）に引き継がれている。

第5節のまとめ英文（The summary of section 5）

①　The organization originally established for providing the maritime communication service, and managed under the international agreement. The extent of the service eventually has been developed to cover aircraft and ground mobile operations using the satellites name of INMARSAT. By the amendment of the agreement , the operational section has been transferred to a private sector (Inmarsat Corp.)

②　GPS has a greater accuracy than the conventional navigation systems and most receivers are capable of simultaneously receiving and using during high-speed transportation anywhere on the earth. There are six orbital planes containing at least four satellites on each and equally spaced 20,000 km above the surface. GPS system consists of 24 satellites or more that orbit the earth and the supporting unit on the ground. When required number of satellites are received, highly accurate position can be obtained anywhere on the earth. The position information provided by GPS has a higher accuracy than of VOR/DME. However, when the signals are used throughout all flight phases, differential GPS is required due to lack of the availability and continuity.

③ Due to interoperability with Inmarsat in terms of aeronautical satellite communication safety service, MTSAT made a contribution as an international safety infrastructure in Asia Pacific region to overseas air traffic control agencies and airlines. It worked for navigation and surveillance as well as communication to provide SBAS (Satellite-Based Augmentation System), which supports safer and securer navigation for oceanic flights, in particular, to improve accuracy and reliability of GPS information, by delivering measurement errors and corrections, and augmentation through satellites. In 2020, its aeronautical satellite solutions were terminated. SBAS function is being succeeded by QZSS (Quasi-Zenith Satellite System).

平成24年3月20日 初版第1刷発行
令和3年3月25日 3版第1刷発行
令和6年4月15日 3版第2刷発行

航空無線通信士

英　語

（電略　コエ）

編集・発行　一般財団法人　情報通信振興会

郵便番号　170−8480
東京都豊島区駒込2−3−10
販売　電話　(03) 3940−3951
FAX　(03) 3940−4055
編集　電話　(03) 3940−8900
振替口座　00100−9−19918
URL　https://www.dsk.or.jp/

印刷所　株式会社エム．ティ．ディ

ISBN978-4-8076-0939-0 C3055 ¥1400E